T0216316

Multiscale and Multiphysics Modeling of Nuclear Facilities with Coupled Codes and its Uncertainty Quantification and Sensitivity Analysis

Chunyu Liu

Multiscale and Multiphysics Modeling of Nuclear Facilities with Coupled Codes and its Uncertainty Quantification and Sensitivity Analysis

Springer Spektrum

Chunyu Liu
Technical University of Munich (TUM)
Munich, Germany

ISBN 978-3-658-43421-2   ISBN 978-3-658-43422-9   (eBook)
https://doi.org/10.1007/978-3-658-43422-9

This Springer Spektrum imprint is published by the registered company Springer Fachmedien
Wiesbaden GmbH, part of Springer Nature.
The registered company address is: Abraham-Lincoln-Str. 46, 65189 Wiesbaden, Germany

Paper in this product is recyclable.

Acknowledgements

This dissertation has been accepted by the School of Engineering and Design of the Technical University of Munich. I would like to thank everyone who helped me.

First of all, I would like to express my deep gratitude to my doctoral father *Prof. Dr. Rafael Macián-Juan* for giving me the opportunity to do my doctoral work and for his personal support since the beginning of my doctoral studies. His immense knowledge and wealth of experience have broadened my horizons and encouraged me throughout all the time of my academic research.

Next, I would also like to thank *Prof. Dr. Antonino Cardella* for his positive expert opinions on the selection of the working fluid in a highly irradiated environment and for his technical support.

I would like to thank my mentor *Dr. Angel Papukchiev* for his technical support and providing the coupling interface between Ansys CFX and ATHLET. I also appreciate the support I received from *Dr. Clotaire Geffray* and *Dr. Run Luo*.

My gratitude extends to the TUM School of Engineering and Design for giving me the opportunity to study at the Chair of Nuclear Technology, Technical University of Munich, and to the European Commission (EC), as this work was partly funded by the SESAME project (grant agreement number 654935). Especially *Petra Popp-Münich* give me a lot of help and encouragement from the beginning of my studies at TUM.

Finally, I would like to thank my parents *Hua Liu* and *Yumei Guo,* my girlfriend *Qiongying Cai,* and my friend *Maomao*. Without their great understanding and encouragement over the past few years, it would be impossible for me to bring this project to a successful conclusion.

Abstract

In today's industries (e.g. nuclear power industry, aerospace industry), various systems are operated in a wide range of regimes. To understand their behaviors, multiscale and multiphysics models have to be developed. Despite the increased modeling accuracy, the simulation results are always subject to uncertainties. Therefore, safety-relevant simulation results should be qualified with uncertainties so that decision makers can make use of them with the necessary confidence. In addition to uncertainty, sensitivity analysis based on statistical techniques can be applied to identify weak points, to quantify the parameter influences on the local/global system behavior, and to support the design and implementation of experiments and measurements involving these systems.

In this dissertation, the application of uncertainty quantification and sensitivity analysis methodology in coupled codes was developed and validated against an experimental loop, and then applied to the design and optimization of a real advanced reactor system with similar working fluid: a single channel model for its local behaviors, and a full core model for its global behaviors.

First, a coupled system and 3D computational fluid dynamics model of the experimental loop TALL-3D, was developed to validate the application of the uncertainty quantification and sensitivity analysis methodology in coupled codes. The results indicate that the uncertainty quantification and sensitivity analysis methodology can be applied to the coupled code system and is capable of providing uncertainty intervals for simulation results that provide more comprehensive information about the real system. The sensitivity analysis allows the determination of the influences of input variables on the output/response, which can be used for the dimension reduction and provides a reference for the experimentalists and manufacturers to reduce the system uncertainty.

Secondly, a coupled 3D neutronics/thermal-hydraulics time-dependent model has been built to investigate the local behavior of a single flow channel of the small modular dual fluid reactor using the validated uncertainty quantification and sensitivity analysis methodology. A strong coupling effect of the neutronics and thermal-hydraulics fields was thoroughly investigated, taking into account of the input uncertainty, under steady and transient scenarios. The results indicate that this dual fluid design is capable to be implemented in an advanced nuclear system and has a high perturbation tolerance.

Finally, a coupled Monte-Carlo and 3D computational fluid dynamics model was established to investigate the global behavior of the entire core. Based on the obtained results, a system model with control system was built and the errors introduced by the dimensionality reduction were quantified by the inverse uncertainty quantification technique. The results show that the designed control system was able to maintain the system stability and regulate the power as expected. Through the uncertainty-based optimization, the delivered control system has the optimized performance: less overshoot, less oscillation, and less time to reach the new steady state.

This work makes a significant contribution to the development, validation and application of various coupling schemes and uncertainty quantification and sensitivity analysis approaches in the modeling of complex nuclear facilities, and provides a reference for further design and safety analysis activities.

Zusammenfassung

In der heutigen Industrie (z. B. Kernkraftwerke, Luft- und Raumfahrtindustrie) werden verschiedene Systeme in einem breiten Spektrum von Regimen betrieben. Um ihr Verhalten zu verstehen, müssen Multiskalen- und Multiphysikmodelle entwickelt werden. Trotz steigender Modellierungsgenauigkeit sind die Simulationsergebnisse immer mit Unsicherheiten behaftet. Sicherheitsrelevante Simulationsergebnisse sollten daher mit Unsicherheiten versehen werden, damit sie von Entscheidungsträgern mit der notwendigen Sicherheit verwendet werden können. Zusätzlich zur Unsicherheit kann eine auf statistischen Techniken basierende Sensitivitätsanalyse eingesetzt werden, um Schwachstellen zu identifizieren, den Einfluss von Parametern auf das lokale/globale Systemverhalten zu quantifizieren und die Planung und Durchführung von Experimenten und Messungen an diesen Systemen zu unterstützen.

In dieser Arbeit wurde eine Methodik zur Quantifizierung von Unsicherheiten und zur Sensitivitätsanalyse in gekoppelten Codes entwickelt und anhand eines experimentellen Kreislaufs validiert. Diese Methodik wurde dann auf den Entwurf und die Optimierung eines realen fortgeschrittenen Reaktorsystems mit einem ähnlichen Arbeitsmedium angewendet: ein Einkanalmodell für sein lokales Verhalten und ein vollständiges Kernmodell für sein globales Verhalten.

Zunächst wurde ein gekoppeltes System- und 3D Computational Fluid Dynamics-Modell des experimentellen Kreislaufs TALL-3D entwickelt, um die Anwendung der Methodik zur Quantifizierung von Unsicherheiten und zur Sensitivitätsanalyse in gekoppelten Codes zu validieren. Die Ergebnisse zeigen, dass die Methodik der Unsicherheitsquantifizierung und Sensitivitätsanalyse auf das gekoppelte Codesystem angewendet werden kann und in der Lage ist, Unsicherheitsintervalle für Simulationsergebnisse zu liefern, die umfassendere Informationen über das reale System liefern. Die Sensitivitätsanalyse ermöglichte die Bestimmung des Einflusses der Eingangsvariablen auf die Ausgang/Reaktion, was zur Dimensionsreduktion genutzt werden kann und eine Referenz für die Experimentatoren und Hersteller zur Reduzierung der Systemunsicherheit darstellt.

Zweitens wurde ein gekoppeltes 3D-Neutronik/Thermohydraulik-Zeitmodell erstellt, um das lokale Verhalten eines einzelnen Strömungskanals des kleinen modularen Dual-Fluid-Reaktors unter Verwendung der validierten Methodik zur Unsicherheitsquantifizierung und Sensitivitätsanalyse zu untersuchen. Ein starker Kopplungseffekt zwischen der Neutronik und der Thermohydraulik wurde unter Berücksichtigung der Eingabeunsicherheiten in stationären und transienten Szenarien detailliert untersucht. Die Ergebnisse zeigen, dass dieses duale Fluid-design in einem fortgeschrittenen Nuklearsystem implementiert werden kann und eine hohe Störungstoleranz aufweist.

Schließlich wurde ein gekoppeltes Monte-Carlo- und 3D Computational Fluid Dynamics-Modell erstellt, um das globale Verhalten des gesamten Kerns zu untersuchen. Basierend auf den erzielten Ergebnissen wurde ein Systemmodell mit Regelungssystem erstellt, und die durch die Dimensionalitätsreduktion einge-führten Fehler wurden mit der Technik der inversen Unsicherheitsquantifizierung quantifiziert. Die Ergebnisse zeigen, dass das entworfene Regelsystem in der Lage war, die Systemstabilität aufrechtzuerhalten und die Leistung wie erwartet zu regeln. Die unsicherheitsbasierte Optimierung führte zu einem optimierten Regelverhalten: weniger Überschwingen, weniger Oszillation und weniger Zeit für den Übergang zum neuen stationären Zustand.

Diese Arbeit leistet einen wichtigen Beitrag zur Entwicklung, Validierung und Anwendung verschiedener Kopplungsschemata und Ansätze zur Quantifizierung von Unsicherheiten und Sensitivitätsanalysen bei der Modellierung komplexer kerntechnischer Anlagen und stellt eine Referenz für weitere Auslegungs- und Sicherheitsanalysen dar.

Contents

Abbrevations

ATHLET   Analysis of THermal-hydraulics of LEaks and Transients
CFD   Computational Fluid Dynamics
DFR   Dual Fluid Reactor
DNP   Delayed Neutron Precursor
DNPs   Delayed Neutron Precursors
ECC   Emergency Core Coolant
FEBE   Forward Euler, Backward Euler
GRS   Gesellschaft für Anlagen- und Reaktorsicherheit (GRS) gGmbH
LBE   Lead-Bismuth Eutectic
LCOs   Limit Cycle Oscillations
LWR   Light Water Reactor
MC   Multi-Component model
ODE   Ordinary Differential Equations
PDF   Probability Distribution function
PDFs   Probability Distribution functions
RANS   Reynolds-Averaged Navier-Stokes
SESAME   thermal-hydraulics Simulations and Experiments for the Safety
          Assessment of MEtal cooled reactors
SMDFR   Small Modular Dual Fluid Reactor
TFD   Thermo-Fluid Dynamics
TFO   Thermo-Fluiddynamic Object
TFOs   Thermo-Fluiddynamic Objects
UQ   Uncertainty quantification

Nomenclature

P_0	Nominal power
z	Axial coordinate, quantile
D	Diameter, hydraulic diameter, diffusion coefficient
H	Height
C_p	Specific heat at constant pressure
k	Thermal conductivity, number of input variables
ρ	Density
μ	Dynamic viscosity, Lagrange multiplier
α	Void fraction
t	Time
w, u	Velocity
ψ	Interphase mass exchange per unit volume
h	Specific enthalpy
p, P	Pressure
τ_i	Interfacial shear per unit volume
g	Gravity constant
q	Heat flux, volumetric flow
S	Momentum source
f_w	wall friction force per unit volume
Λ	Mean neutron generation time
β_i	Fraction of the i-th DNP group
β	Overall fraction of the DNP groups
λ	Decay constant
ϕ_{tot}	Total neutron flux
$\frac{1}{v}$	Inverse of the neutron velocity
N_0	Initial neutron density

τ_c	Transit time inside the core
τ_e	Transit time outside of the core
λ_c	Drifting-induced decay constant inside the core
λ_e	Drifting-induced decay constant outside of the core
α_f	Reactivity feedback coefficient of fuel
α_c	Reactivity feedback coefficient of coolant
ρ_0	Initial reactivity
ρ_{insert}	Inserted reactivity
$C_{i,0}$	Initial concentration of the i-th DNP group inside the core
$Ce_{i,0}$	Initial concentration of the i-th DNP group outside of the core
A_{fw}	Heat transfer area between fuel and pipe wall
A_{wc}	Heat transfer area between coolant and pipe wall
$\dot{m}_f$	Mass flow rate of fuel
$\dot{m}_c$	Mass flow rate of coolant
M_f	Mass of fuel
M_w	Mass of pipe wall
M_c	Mass of coolant
$c_{p,f}$	Heat capacity of fuel
$c_{p,w}$	Heat capacity of pipe wall
$c_{p,c}$	Heat capacity of coolant
h_{fw}	Heat transfer coefficient between fuel and pipe wall
h_{wc}	Heat transfer coefficient between coolant and pipe wall
T_f	Fuel temperature, fluid temperature
T_w	Pipe wall temperature
T_c	Coolant temperature
T_f^{in}	Fuel inlet temperature
T_c^{in}	Coolant inlet temperature
$e(t)$	Difference between the set value and the real (measured) value
$u(t)$	Calculated controller output
K_p	Proportional gain
K_i	Integral gain
v	Neutron speed
Φ	Neutron flux
Σ	Macroscopic cross section
χ_p^g	Fraction of the emitted prompt neutrons in group g
χ_d^g	Fraction of the emitted delayed neutrons in group g
ν	Number of neutrons emitted in fission
C	Concentration of the delayed neutron precursors
$\mathbf{u}$	Average velocity field

$\mathbf{I}$	Identity matrix
$\mathbf{K}$	Viscosity term
$\mathbf{F}$	Body force vector
μ_T	Turbulent viscosity
k	Turbulent kinetic energy
ϵ	Turbulent dissipation rate
$C_{\epsilon 1}, C_{\epsilon 2}, C_\mu$	Model constant
$\sigma_k, \sigma_\epsilon$	Model constant
$\mathbf{q}$	Heat flux by conduction
$\mathbf{n}$	Normal vector
u_τ	Friction velocity
T_+	Dimensionless temperature
Pr_T	Turbulent Prandtl number
$E()$	Expectation
y, Y	Dependent variable
N	Sample size
σ	Standard deviation
σ^2	Variance
$\bar{y}$	Arithmetic mean of y
$Var()$	Variance
$1 - \alpha$	Confidence level
p	Population proportion
CI	Confidence interval
$r, N - s + 1$	Index of tolerance limit in the ordered sequence of output value
r, R	Standard regression coefficient
R^2	Coefficient of multiple determination
$Corr()$	Correlation
f	A constant or function
S	Main effect index
S_T	Total effect index
ω, W	Weight
γ	Semivariogram
σ_K^2	Kriging error
h_w	Energy loss
$ITAE$	Integral time-weighted absolute error
IAE	Integral absolute error

List of Figures

List of Tables

Introduction

1

In today's industries (e.g. nuclear power industry, aerospace industry), there are various systems operating in a wide range of regimes. Since it is essential for the engineers to understand their behaviors in both steady and transient scenarios, multiphysics and multiscale modeling of these systems attracts more and more attention. Despite the increased modeling accuracy achieved by advanced approaches, the simulation results are always subject to uncertainty. Therefore, safety relevant simulation results should be qualified with uncertainty, so that decision makers can make use of them with the necessary confidence. In addition to uncertainty, sensitivity analysis based on statistical techniques can be applied to identify weak points in the system design, to quantify the influences of local and global parameters on the local/global system behavior, and to support the design and implementation of experiments and measurements.

This chapter is organized as follows. The first section (Section 1.1) describes the basic terms: multiphysics modeling, multiscale modeling, uncertainty quantification. The thermal-hydraulics Simulations and Experiments for the Safety Assessment of MEtal cooled reactors (SESAME) project and the Dual Fluid Reactor (DFR) concept are introduced in Section 1.2 and Section 1.3, respectively. Then the state of the art of uncertainty quantification is reviewed in Section 1.4. Finally the objective of this study is described in Section 1.5 and the outline of this dissertation is given in Section 1.6.

© The Author(s), under exclusive license to Springer Fachmedien Wiesbaden GmbH, part of Springer Nature 2023
C. Liu, *Multiscale and Multiphysics Modeling of Nuclear Facilities with Coupled Codes and its Uncertainty Quantification and Sensitivity Analysis*,
https://doi.org/10.1007/978-3-658-43422-9_1

1.1   Introduction to Basic Terms

1.1.1   Multiphysics Modeling

In the primary circuit of the nuclear reactors, the nuclear chain reaction and heat transfer take place simultaneously. Thus, a strong coupling effect of the neutronics and thermal-hydraulic fields is thus expected and needs to be thoroughly investigated under both steady and transient scenarios. It is very important to develop multiphysics modeling approaches to couple the different physical fields internally or externally. The simulations based on these approaches provide a good representation of the individual physical fields, and also the coupling effect between them. Thus, the interaction and feedback between these physical fields can be quantified and a high fidelity system response can be provided.

1.1.2   Multiscale Modeling

Many thermal-hydraulic systems are quite complex and time-consuming to simulate numerically. Therefore, it is very important to develop fast 3D approaches for the detailed analysis of these complex thermal-hydraulic systems. The simulations based on these approaches provide a good representation of the detailed thermal-hydraulic behavior through CFD calculations in the regions where small-scale knowledge is needed to better represent the macroscopic system responses. For those regions where the macroscopic thermal-hydraulic descriptions are sufficient to represent their influence on the system responses, a coarser resolution allows for significant savings of computational resources.

1.1.3   Uncertainty Quantification

Regardless of the accuracy of the computational methodology, a degree of uncertainty is always built into the process due to the approximations and assumptions required to produce workable simulation models and calculations. Therefore, safety-related results should be accompanied by an estimate of the impact of modeling uncertainty on the simulation results so that those making decisions based on them can do so with the necessary confidence. Sensitivity analysis based on statistical techniques is a powerful tool for the identification of weak points in the design, local and system-wide parameter influences on thermal-hydraulics behaviors, etc.

It can also be a very helpful tool in the design of experiments and in the evaluation of their results.

1.2 Introduction to SESAME Project

The SESAME project supports the development of European liquid metal cooled reactors by coordinating a series of thermal-hydraulics simulations and experiments among its 23 partners. [SES] In the framework of Work Package 5 (WP5 Integral System Simulations), a specific activity "the uncertainty and sensitivity analysis of the TALL-3D facility" has been assigned to implement coupled simulations in a massively parallel environment and of statistical techniques to perform the uncertainty quantification and sensitivity analysis for the thermal-hydraulics of the liquid metal fast reactor. The selected exercise was based on the experimental campaign carried out on the TALL-3D facility. The experimental campaign, carried out in the framework of SESAME WP4 (Integral System Reference Data), consisted of 3 test groups: test group 01 for the measurement uncertainty quantification, test group 02 for the code input calibration and test group 03 for calibration, benchmark and code validation. Among these tests, TG03.S301.03 was selected and in this dissertation the uncertainty and sensitivity analysis based on the coupled ATHLET-ANSYS CFX code system focuses on this case.

This dissertation reports the uncertainty and sensitivity analysis results of the benchmark test TG03.S301.03 based on the numerical simulations using coupled codes and standalone system code, as a semi-blind case and as an open case, respectively.

In order to investigate the thermal-hydraulic behaviors of Lead-Bismuth Eutectic (LBE), an experimental facility named TALL-3D has been constructed at the Royal Institute of Technology (KTH). The setup and the experimental campaign of this facility are briefly described in the following sections. A more detailed and comprehensive description can be found in the SESAME deliverable: D3.1 TALL-3D setup.

1.2.1 Setup of Tall-3D Facility

The TALL-3D experimental facility is about 6.5 m high and consists of two circuits: the primary circuit and the secondary circuit. (Figure 1.1)[GJK+15]

The primary circuit (Figure 1.2[PGGK16], left) uses LBE as the working fluid and consists of 3 vertical legs.

- The left leg is called the Main Heater (MH) leg. A heated pin is immersed in the lower part of this leg and is used to heat the LBE fluid in the left leg.
- The middle leg is called the Test Section (TS) leg. In the lower part of this leg a 3D test section (Figure 1.2[PGGK16], middle) is installed, where the complex 3D thermal-hydraulic phenomena (such as LBE mixing and stratification) are simulated by CFX code (ANSYS CFX) and investigated by the numerical simulation, in order to validate the code capabilities. More than 140 thermocouples (TCs) are installed In the TS to monitor the temperature distribution. The main TC groups in the test section are (Figure 1.2[PGGK16], right): CIP (circular inner plate TCs), BP (bottom plate TCs), IPT (inner pipe TCs), ILW (inner lateral wall TCs), OLW (outer lateral wall TCs). Each TC group is found four times in the cylindrical pool - at $0°$, $90°$, $180°$ and $270°$. Since the simulated transients are symmetric, the average of the 4 TCs is compared with the numerical prediction. The uncertainty of TC measurement is 2.2 K.[GJK+15]
- The right leg is called the Heat Exchanger (HX) leg. In the upper part of this leg a heat exchanger is installed, which is intended to remove the heat produced by MH and TS from the primary circuit.

The secondary circuit uses oil as working fluid and is connected to the primary circuit by the HX. In this dissertation the secondary side only acts as a heat sink for the primary circuit.

1.2.2   Experimental Campaign of Tall-3D Facility

In order to investigate the system behavior during the transition from forced to natural convection and from natural to forced convection in the primary circuit, a series of transient tests were performed during which dynamic thermal-hydraulic phenomena occurred before the final steady state was reached. The main initial and final conditions for all the transient tests in test group 03 are summarized in Table 1.1.

Three of these tests were selected for the benchmark activities: TG03.S301.03, TG03.S301.04, TG03.S310.01. For TG03.S301.03 and TG03.S301.04, the forced to natural transient, both MH and 3D TS heaters were operated at constant power. While for TG03.S310.01, the forced to natural transient with MH at constant power and 3D TS heater kept off at the beginning of pump trip.

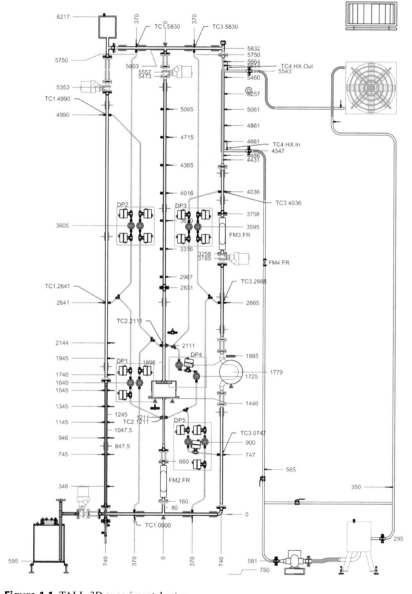

Figure 1.1 TALL-3D experimental setup

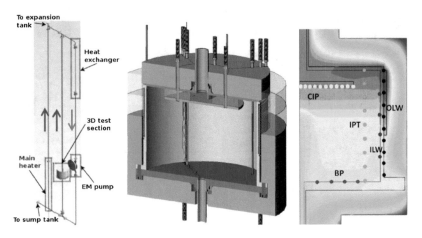

Figure 1.2 TALL-3D primary circuit (left), test section (middle) and position of the TC groups (right)

1.3   Introduction to DFR Concept

The study of the two-fluid molten salt reactor can be traced back to 1966, when a two-fluid molten salt breeder reactor (MSBR) was designed at the Oak Ridge National Laboratory (ORNL) [RBSB70]. The load-following capability of this MSBR system was studied for various ramp rates of the power demand [SLW+18]. Later a fast breeder reactor of about 2 GW thermal power using molten chlorides as both fuel and coolant was proposed [TL74]. The Dual Fluid Reactor (DFR) concept, as a variant of the molten salt reactor, was proposed by the researchers at the Institute for Solid-State Nuclear Physics (IFK). In this reactor concept, molten salt is used as fuel undergoing fission in the primary circuit, which is the same as in a normal molten salt reactor, while liquid lead is used as coolant in the secondary circuit. As a fast spectrum reactor, graphite is not required and the heat transfer between the molten salt and liquid lead takes place in the core. With the advantages of both molten salt and lead-cooled fast reactors, the dual fluid reactor concept is able to meet the Generation IV International Forum [GIF17] criteria for reactors. A series of studies covering the neutronic characteristics, sensitivity performance, coupled computation, estimation of the emergency drain tank, depletion as well as thermal-hydraulic simulations can be found in references [WMJS15, Wan17, WMJ17a, WMJ17b, WMJ18, WLMJ19, LLLMJ20].

Table 1.1 Transient Tests

#	Name	HX mass flow rate [kg/s]	TC3.5830 [°C]	MH Power [kW]	TS Power [kW]
1	TG03.S301.01	4.6→0.5	245→270	3.2→3.2	5.5→5.5
2	TG03.S301.02	4.6→0.5	227→245	2.5→2.5	4.9→4.9
3	TG03.S301.03	4.7→xxx	288→xxx	0.72→0.72	10.3→10.3
4	TG03.S301.04	3.3→0.5	250→280	3.2→3.2	4.0→4.0
5	TG03.S301.05	3.3→0.53	250→284	3.2→3.2	4.2→4.2
6	TG03.S301.06	4.8→0.59	292→335	1.1→1.1	10.6→10.6
7	TG03.S301.07	4.8→0.3	241→255	0.45→0.45	5.6→5.6
8	TG03.S302.01	4.3→0.5	182→271	2.3→2.3	0→4.9
9	TG03.S306.01	0.5→4.6	244→N/A	2.5→2.5	4.8→4.8
10	TG03.S307.01	0.5→4.6	269→240	3.2→8.6	5.5→0
11	TG03.S310.01	4.6→0.5	241→274	8.6→8.6	0→0
12	TG03.S311.01	0.5→0.45	280→290	3.2→0.6	4.0→6.7

In order to develop such a novel reactor, the thermal-hydraulic and neutronics characteristics, as well as their coupling effects, have to be thoroughly investigated. In this dissertation, a single channel model and a full core model were established, using explicit and implicit coupling schemes, respectively.

In contrast to the original dual fluid reactor design, in this dissertation the nominal power of the core was reduced to: $P_0 = 0.1$ GW $= 100$ MW , which is much lower compared to the originally designed thermal power of 3 GW, and is named as Small Modular Dual Fluid Reactor (SMDFR). The schematic of the SMDFR is shown in Figure 1.3. There are three circuits in the schematic: the primary circuit of molten salt as fuel is shown in brown; the secondary circuit of liquid lead as the primary coolant is shown in yellow; and the tertiary circuit of liquid lead as the secondary coolant is shown in blue. The mass flow rate of the molten salt in the primary circuit is very low and is used solely for online fuel processing rather than heat removal. The heat generated by the fission reaction in the core is mainly absorbed by the liquid lead as the primary coolant in the secondary circuit and then transferred to the secondary coolant in the tertiary circuit for utilization, such as hydrogen production, electricity generation, and water desalination.

In this dissertation, the core of the SMDFR is chosen as the object of research, which consists of three parts: the distribution zone, the core zone and the collection

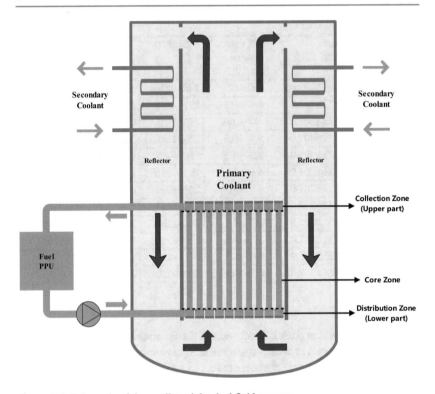

Figure 1.3 Schematic of the small modular dual fluid reactor

zone. Starting from the lower part of the core, the fuel (molten salt) enters the distribution zone, flows through the core zone, and finally leaves the collection zone. The coolant (liquid lead) flows in the same direction and absorbs the heat generated by the fission reaction taking place in the molten salt.

The main design parameters are listed in Table 1.2. In this dissertation, the molten salt, which is a mixture of uranium tetrachloride and plutonium tetrachloride, is selected as the fuel, and silicon carbide is used as the pipe wall.

Table 1.2 SMDFR design parameters

Parameters	Values
Core zone D × H (m)	0.95 × 2.0
Distribution zone D × H (m)	0.95 × 0.2
Collection zone D × H (m)	0.95 × 0.2
Height of core (m)	2.4
Outer reflector diameter (m)	1.25
Tank D × H (m)	1.65 × 3.4
Number of fuel tubes	1027
Fuel pin pitch (m)	0.025
Outer/interior fuel tube diameter (m)	0.008/0.007
Outer/interior coolant tube diameter (m)	0.005/0.004
Mean linear power density (W/cm)	609
Fuel inlet/outlet temperature (K)	1300/1300
Coolant inlet/outlet temperature (K)	973/1100
Fuel inlet/in-core velocity (m/s)	3/0.5225
Coolant inlet/in-core velocity (m/s)	5/1.3488

The thermophysical properties of the materials are listed in Table 1.3 with reference to the literatures [DKRC75][NMH+97][FSA+15][HRW+17]. The formulas are applied for each 100 K interval from 800K to 1600K, which fully covers the operating temperature range. The corresponding interpolated values based on these temperature points are used for the calculation.

Table 1.3 Thermophysical properties

	Formula	Validity range [K]
Fuel specific heat capacity C_p (g/cm^3)	400	[–]
Fuel thermal conductivity k $(W/(m \cdot K))$	2	[–]
Fuel density ρ (kg/m^3)	$1000 \times (5.601 - 1.5222 \times 10^{-3} \times T)$	[–]
Fuel dynamic viscosity μ $(Pa \cdot s)$	4.50×10^{-4}	[–]
Coolant specific heat capacity C_p (g/cm^3)	$176.2 - 0.04923 \times T + 1.544 \times 10^{-5} \times T^2 - 1.524 \times 10^6 * T^{-2}$	[600–1500]
Coolant thermal conductivity k $(W/(m \cdot K))$	$9.2 + 0.011 \times T$	[600–1300]
Coolant density ρ (kg/m^3)	$11441 - 1.2795 \times T$	[600–2000]
Coolant dynamic viscosity μ $(Pa \cdot s)$	$4.55 \times 10^{-4} \times e^{1069/K}$	[600–1473]
Pipe wall heat capacity C_p (g/cm^3)	690	[–]
Pipe wall conductivity k $(W/(m \cdot K))$	$61100/(T - 115)$	[300–2300]
Pipe wall density ρ (kg/m^3)	3210	[–]

1.3.1   Geometry of the Distribution Zone Model of the SMDFR

The arrangement of fuel and coolant tubes in the SMDFR core is shown in Figure 1.4. A complete calculation of the core would require very large computational resources. Due to the symmetric hexagonal configuration of the core region, a sector of 30° (Figure 1.5, right) of the lower part of the core with a height of 0.4 m (from z = 0 m to z = 0.4 m, Figure 1.5, left) around the central cylindrical axis, only one twelfth of the full geometry, was selected to represent the full-scale geometry by applying symmetric boundary conditions (mirror plane) [WLMJ19]. The number of fuel tubes to be simulated was then reduced from 1027 to 100, including 72 complete tubes. A complete tube means that the whole tube was included in the selected 30° sector and an incomplete tube means that only a part of the tube was included. The tubes at the boundary were incomplete tubes, as shown in Figure 1.5 (right).

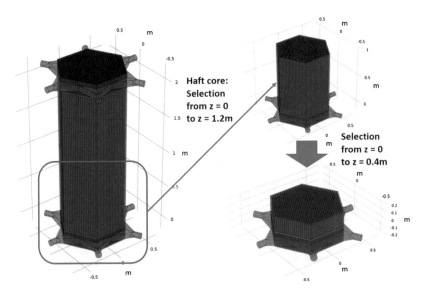

Figure 1.4 The arrangement of both fuel and coolant tubes in the SMDFR core: left, full height, z = 0 m to z = 2.4 m; upper right, half height, z = 0 m to z = 1.2 m; lower right, z = 0 m to z = 0.4 m

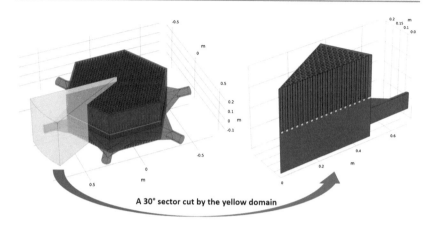

A 30° sector cut by the yellow domain

Figure 1.5 A 30° sector of the distribution zone with a height of 0.4 m, z = 0 m to z = 0.4 m

In Figure 1.5 (right), three colored domains can be observed: brown, red, and gray, representing the fuel domain, the coolant domain, and the pipe wall domain, respectively. This geometric model contained the distribution zone with a height of 0.2 m and its upper part (starting from the core zone) with a height of 0.2 m. In the fuel domain, the molten salt entered through the inlet pipe from the right side, flowed into the distribution zone, spread in the gaps between the hexagonally arranged coolant tubes, flowed upward into the fuel tubes, and then left the core zone through the upper part. For the coolant domain, the liquid lead flowed through the lower coolant pipes into the core area, flowed upwards between the gaps along the fuel tubes, and, finally, left the core area through the upper part.

1.3.2 Geometry of the Single Channel Model of the SMDFR

As shown in Figure 1.6, a single channel of the full core was selected to investigate the system characteristics and the coupling effect of the two fluids. This geometry was set up in COMSOL Multiphysics for both the thermal-hydraulic and neutronics fields. These two physical fields are implicitly coupled to predict the system behaviors under steady state and transient scenarios.

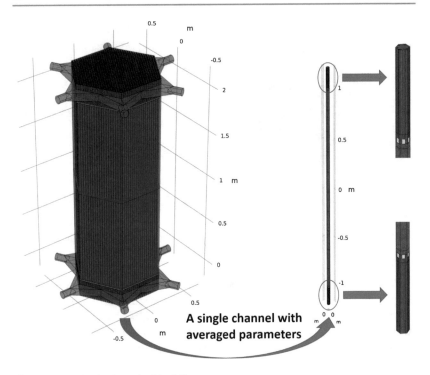

Figure 1.6 A single channel of the full core

1.3.3   Geometry of the Full Core Model of the SMDFR

As shown in Figure 1.7, a 30° sector of the full core was selected to represent
the full-scale geometry by applying symmetric boundary conditions. This geometry was set up in both the COMSOL Multiphysics and the Serpent codes for the
thermal-hydraulic and neutronics fields, respectively. The two physical fields are
then externally coupled to predict the steady-state characteristics of the SMDFR
core.

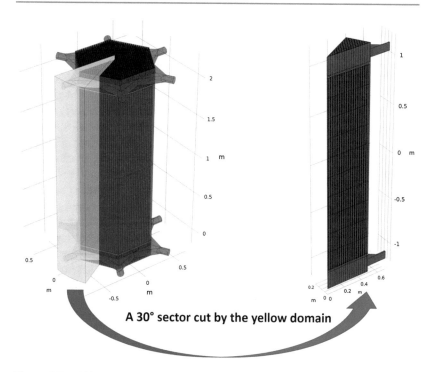

A 30° sector cut by the yellow domain

Figure 1.7 A 30° sector of the full core

1.4   State of the Art of Uncertainty Quantification

In the work of Wicaksono [Wic18], a methodology has been developed to quantify the uncertainty of physical model parameters implemented in a thermal-hydraulic system code based on the experimental data. This methodology is based on a listic framework and consists of three steps: global sensitivity analysis, metamodeling and Bayesian calibration.

As the development of the best-estimate codes, the evaluation of the margin to acceptance criteria should be based on the upper limit of the calculated uncertainty range. Instead of computing a statistical quantile (e.g. 95%) and the associated confidence interval, which requires runs of a few hundred to a few thousand, Glaeser [Gla08] proposed tolerance limits to quantify the uncertainty of computer code results with a series of runs, the number of which can be reduced to less than one hundred.

A methodology was proposed by Pourgol-Mohamad et al. [PMMM09] to make thermal-hydraulic system codes capable of comprehensively supporting the uncertainty assessment with the ability to handle important accident transients by using the Bayesian approach for incorporating available evidence in quantifying uncertainties in the code predictions. Later, Pourgol-Mohamad et al. [PMMM10] proposed a methodology for updating the results of uncertainty quantification of thermal-hydraulic system codes when additional data from tests and experiments become available.

The work of Cho et al. [CSLM17] proposed an effective approach to quantify uncertainties associated with transient simulation results of a dynamic solar thermal energy system model with uncertain parameters. Using the concept of impulse response and convolution, the sensitivities of time-varying external inputs were estimated and the number of required simulations can be significantly reduced.

In addition to the thermal-hydraulic codes, uncertainty also exists in the results of Monte Carlo codes. In the coupled Serpent/RELAP5 model, Wu et al. [WK14] presented the results of the investigation of how the statistical uncertainty in Monte Carlo codes affects the simulation results when thermal-hydraulic feedback is introduced.

In addition to the statistical uncertainty, there are uncertainties due to the initial and boundary conditions, the physical parameters and some physical models. In the modern nuclear industry, the application of coupled multiscale and multiphysics codes (coupling of neutronics and thermal-hydraulic effects) instead of a standalone code for safety analysis is attracting more and more attention. The quantification of the effect of these uncertainties in the multiscale and multiphysics calculations is very important for the evaluation of the calculation results.

As a step towards validating the application of coupled codes for the simulation of complex flows, uncertainty propagation is applied to multiscale thermal-hydraulic coupled codes, was presented by Geffray [Gef17].

1.5   Objective

The main objectives of this work are:

- Performing uncertainty quantification and sensitivity analysis based on the coupled system and 3D computational fluid dynamics model of the TALL-3D facility;

- Validation of the application of the uncertainty quantification and sensitivity analysis methodology in coupled codes by the experimental data;
- The development of implicit/explicit coupling schemes and their application in the modeling of the SMDFR core;
- The application of the validated uncertainty quantification and sensitivity analysis methodology to multiphysics models of the SMDFR core for safety analysis;
- Identification of key parameters influencing the system behavior;
- Concluding the qualification of the uncertainty quantification and sensitivity analysis methodology for application to coupled models of complex nuclear facilities.

1.6 Outline

Outline of the dissertation:

Chap. 1 gives an introduction to: basic terms, the SMDFR, the SESAME project, the state of the art of Uncertainty quantification (UQ), objective, and outline.

Chap. 2 presents the methodology of multiscale and multiphysics modeling, and the methodology of uncertainty quantification and sensitivity analysis applied in coupled codes is described.

Chap. 3 shows how to verify that the computational models are sufficiently accurate to predict the system behavior.

Chap. 4 presents the results obtained by the coupled system and 3D computational fluid dynamics model of TG03.S301.03 in the SESAME project, which was developed to validate the application of the uncertainty quantification and sensitivity analysis methodology in coupled codes.

Chap. 5 presents the results delivered by the coupled 3D neutronics / thermal-hydraulics time-dependent model, which was built to investigate the steady and transient behaviors of a single flow channel of the SMDFR core applying the validated uncertainty quantification and sensitivity analysis methodology.

Chap. 6 presents the results obtained by the coupled Monte-Carlo / 3D computational fluid dynamics model, which was built to investigate the global
behaviors of the entire core. In addition, a system model based on the
inverse uncertainty propagation technique was built and an uncertainty-
based control scheme was designed and optimized.

Chap. 7 summarizes the main achievements and draws the conclusion of this work.
An outlook on future research is given.

Methodology

2

To achieve the objectives stated in Section 1.5, the modeling and coupling methodology has to be developed and applied, and then the uncertainty quantification and sensitivity analysis can be performed.

This chapter is organized as follows. Section 2.1 describes the modeling methods of various codes for different scales and physical fields. Then the coupling approaches for the SESAME project and for the SMDFR core are introduced in Section 2.2 and Section 2.3, respectively. Finally the uncertainty quantification and sensitivity analysis methods are summarized in Section 2.4.

2.1 Modeling Methods

In this dissertation, five codes have been employed for modeling: Analysis of THermalhydraulics of LEaks and Transients (ATHLET), Serpent, Ansys CFX, COMSOL Multiphysics, MATLAB. The modeling processes can be divided into three categories based on their characteristics: system modeling: ATHLET (Section 2.1.1) and MATLAB (Section 2.1.2), neutron transport modeling: Serpent (Section 2.1.3) and COMSOL Multiphysics (Section 2.1.4), Computational Fluid Dynamics (CFD) modeling: Ansys CFX and COMSOL Multiphysics (Section 2.1.5).

2.1.1 System Modeling in System Analysis Code ATHLET

In this dissertation ATHLET is employed as a system analysis code to perform the numerical simulation of the primary circuit. It is important to understand the

C. Liu, *Multiscale and Multiphysics Modeling of Nuclear Facilities with Coupled Codes and its Uncertainty Quantification and Sensitivity Analysis*, https://doi.org/10.1007/978-3-658-43422-9_2

principles of this code before using it, in order to create an optimal input file and to make a scientific comment on the simulation results. Its overview and the thermo-fluid dynamic model are briefly described in the following sections. The modeling process in ATHLET is presented in detail together with the Ansys CFX model in Section 2.2.

Overview of ATHLET

The thermal-hydraulic computer code ATHLET is being developed by Gesellschaft für Anlagen- und Reaktorsicherheit (GRS) gGmbH (GRS). This code can analyze the operating conditions, abnormal transients and all types of leaks and breaks in nuclear power plants. The main features of the code [ea16a] are:

- advanced thermal-hydraulic modeling (compressible fluids, mechanical and thermal nonequilibrium of vapor and liquid phase)
- divers fluids to be available: light or heavy water, helium, sodium, lead or lead-bismuth eutectic
- heat generation, heat conduction and heat transfer to single- or two-phase fluid considering
- structures of different geometry, e.g. rod or pebble bed
- interfaces to specialized numerical models such as 3D neutron kinetic codes or 3D CFD
- codes for coupled multiphysical or multiscale simulations
- control of ATHLET calculation by callbacks to programming language independent user
- code enabling the coupling of external models
- plug-in technique for user provided code extensions
- modular code architecture
- separation between physical models and numerical methods
- numerous pre- and post-processing tools
- portability
- continuous and comprehensive code validation

ATHLET is being applied by numerous institutions in Germany and abroad.

Range of Applicability

ATHLET has been developed and validated to be used for all types of design basis and beyond design basis incidents and accidents without core damage in light water reactors. The range of working fluids covers light and heavy water, allowing the transition between subcritical and supercritical fluid states. In addition, other coolants can be simulated as working fluids: helium, liquid sodium, lead and lead-bismuth eutectic. [ea16a]

Code Structure

ATHLET is written in FORTRAN. The code structure is highly modular so it allows an easy implementation of different physical models. The code is composed of several basic modules [ea16a] for the calculation of different thermal-hydraulic phenomena emphasizing the operation of a nuclear power reactor:

* Thermo-Fluid Dynamics (TFD)
* Heat Conduction and Heat Transfer (HECU)
* Neutron Kinetics (NEUKIN)
* Control and Balance of Plant (GCSM)

The TFD system of ordinary differential equations is solved fully implicitly using the numerical integration method: Forward Euler, Backward Euler (FEBE).

Fluiddynamics

The Thermo-FluidDynamics (TFD) module of ATHLET employs a modular network approach for the representation of a thermal-hydraulic system. A given system configuration can be reproduced just by connecting basic fluiddynamic elements, called Thermo-Fluiddynamic Objects (TFOs). There are several Thermo-Fluiddynamic Object (TFO) types, each of them is applied with a selected fluiddynamic model. All object types are classified into three basic categories: [ea16a]

* **Pipe objects** employ a one-dimensional TFD Model describing the transport of fluid. After nodalization according to the input data, a pipe object can be understood as a number of consecutive nodes (control volumes) connected by flow paths (junctions). A special application of a pipe object, called single junction pipe, consists of only one junction, without any control volume.
* **Branch objects** consist of only one control volume. They employ a zero-dimensional TFD Model of non-linear ordinary differential equations or algebraic equations.
* **Special objects** are used for network components that exhibit a complex geometry, e.g. the cross connection of pipe objects aligned in parallel for the generation of a multidimensional network.

ATHLET offers two different sets of model equations [ea16a] for the simulation of the fluid-dynamic behavior:

- The **5-equation model** with separate conservation equations for liquid and vapor mass and energy, supplemented by a mixture momentum equation. It accounts for thermal and mechanical non-equilibrium and includes a mixture level tracking capability.
- The **Two-fluid model** with fully phase-separated conservation equations for liquid and vapor mass, energy, and momentum (without mixture level tracking capability). The two-fluid-model is currently the accepted state-of-the-art mode for the simulations involving two-phase flow applications.

The spatial discretization is performed on the basis of a finite-volume staggered-grid approach. The mass and energy equations are solved within control volumes, and the momentum equations are solved over junctions connecting the centers of the control volumes.The solution variables are the pressure, vapor temperature, liquid temperature and vapor mass quality within a control volume, as well as the mass flow rate (5-eq. model) or the phase mass velocities (6-eq. model) in a junction, respectively.

In addition, both fluiddynamic options allow for the simulation of **non-condensable gases**. This applies for water as well as for the liquid metal working fluid. Fluid properties are provided for hydrogen, nitrogen, air, helium and argon. Additional mass conservation equations can be included for the description of boric acid or zinc borate transport within a coolant system as well of the transport and release of nitrogen dissolved in the liquid phase of the coolant. [ea16a]

Both the 5-equation model and the two-fluid model traditionally employ the one-dimensional conservation equations for mass, momentum and energy. An ATHLET multi-dimensional flow model is currently in the validation process.

Numerical Methods

The time integration of the thermo-fluiddynamic model is performed with the general purpose Ordinary Differential Equations (ODE)-solver: FEBE. It provides the solution of a linear system of ODE of first order, splitting it into two subsystems, the first being integrated explicitly, the second implicitly. Generally, the fully implicit option is used in ATHLET. Each TFO provides a subset of the entire ODE system, which is integrated simultaneously by FEBE. [ea16a]

Validation

The development of ATHLET was and is accompanied by a systematic and comprehensive validation program. The validation is mainly based on pre- and post-test calculations of separate effects tests, integral system tests including the major International Standard Problems, as well as on real plant transients. A well balanced set of tests has been derived from the CSNI Code Validations Matrix emphasizing the German combined Emergency Core Coolant (ECC) injection system. The tests cover phenomena which are expected to be relevant for all types of events of the envisaged ATHLET range of application for all common Light Water Reactor (LWR)s including advanced reactor designs with up-to-date passive safety systems. The validation of ATHLET in the field of future Gen IV reactors is underway. [ea16a]

Thermo-Fluid Dynamic Model of ATHLET

Although ATHLET includes several kinds of modules, only the thermo-fluid dynamic module is the emphasis of the numerical simulation of water hammer in this thesis. Besides this, nitrogen has also to be included by implementing Multi-Component model (MC), since it is employed in the experiments as residual gas and pressurization gas. The governing equations and MC are decribed hereinafter. [ea16b]

Basic Field Equations

The ATHLET thermo-fluid dynamic model allows the simulation of the thermal- and fluid dynamic processes of several fluids (working fluids) within a highly flexible flow system defined by the user via input data. Nevertheless, a mixture with non-condensable gases may be modeled which are denoted as 'gas'.

For all working fluid, the system of differential equations used in ATHLET is based on the following general conservation equations for the liquid and vapor phases:

liquid mass

$$\frac{\partial((1-\alpha)\rho_L)}{\partial t} + \nabla \cdot ((1-a)\rho_L \vec{w}_L) = -\psi \tag{2.1}$$

vapor mass

$$\frac{\partial(\alpha\rho_V)}{\partial t} + \nabla \cdot (\alpha\rho_V \vec{w}_V) = \psi \tag{2.2}$$

liquid energy

$$\frac{\partial[(1-\alpha)\rho_L(h_L + \frac{1}{2}\vec{w}_L\vec{w}_L - \frac{p}{\rho_L})]}{\partial t} + \nabla \cdot [(1-\alpha)\rho_L\vec{w}_L(h_L + \frac{1}{2}\vec{w}_L\vec{w}_L)] =$$

(2.3)

$$- p\frac{\partial(1-\alpha)}{\partial t}$$

$+ \vec{\tau}_i\vec{w}_L$	(shear work at the phase interface)
$+ (1-\alpha)\vec{\tau}_i(\vec{w}_V - \vec{w}_L)$	(dissipation due to interfacial shear)
$+ (1-\alpha)\rho_L\vec{g}\vec{w}_L$	(gravitational work)
$+ \dot{q}_{WL}$	(heat flow through structures)
$+ \dot{q}_i$	(heat flow at the phase interface)
$+ \psi(h_{\psi,L} + \frac{1}{2}\vec{w}_\psi\vec{w}_\psi)$	(energy flow due to phase change)
$+ S_{E,L}$	(external source terms)

where:

$\vec{w}_\psi = \vec{w}_L$	(for evaporation)
$\vec{w}_\psi = \vec{w}_V$	(for condensation)

vapor energy

$$\frac{\partial[\alpha\rho_V(h_V + \frac{1}{2}\vec{w}_V\vec{w}_V - \frac{p}{\rho_V})]}{\partial t} + \nabla \cdot [\alpha\rho_V\vec{w}_V(h_V + \frac{1}{2}\vec{w}_V\vec{w}_V)] =$$ (2.4)

$$- p\frac{\partial\alpha}{\partial t}$$

$- \vec{\tau}_i\vec{w}_V$	(shear work at the phase interface)
$+ \alpha\vec{\tau}_i(\vec{w}_V - \vec{w}_L)$	(dissipation due to interfacial shear)
$+ \alpha\rho_V\vec{g}\vec{w}_V$	(gravitational work)
$+ \dot{q}_{WV}$	(heat flow through structures)
$+ \dot{q}_i$	(heat flow at the phase interface)
$+ \psi(h_{\psi,V} + \frac{1}{2}\vec{w}_\psi\vec{w}_\psi)$	(energy flow due to phase change)
$+ S_{E,V}$	(external source terms)

where:

$$\vec{w}_\psi = \vec{w}_L \qquad \text{(for evaporation)}$$
$$\vec{w}_\psi = \vec{w}_V \qquad \text{(for condensation)}$$

liquid momentum

$$\frac{\partial((1-\alpha)\rho_L \vec{w}_L)}{\partial t} + \nabla \cdot ((1-\alpha)\rho_L \vec{w}_L \vec{w}_L) + \nabla((1-\alpha)p) = \qquad (2.5)$$

$$+ \vec{\tau}_i \qquad \qquad \text{(interfacial friction)}$$
$$- (1-\alpha)\vec{f}_W \qquad \qquad \text{(wall friction)}$$
$$- \psi \vec{w}_\Gamma \qquad \qquad \text{(momentum flux due to phase change)}$$
$$- (1-\alpha)\rho_L \vec{g} \qquad \qquad \text{(gravitation)}$$
$$+ \alpha(1-\alpha)(\rho_L - \rho_V)\vec{g}D_h \nabla\alpha \qquad \qquad \text{(water level force)}$$
$$+ \alpha(1-\alpha)\rho_m \left(\frac{\partial \vec{w}_R}{\partial t} + \nabla \vec{w}_R\right) \qquad \qquad \text{(virtual mass)}$$
$$+ S_{I,L} \qquad \qquad \text{(external momentum source terms (e.g. pumps))}$$

vapor momentum

$$\frac{\partial(\alpha\rho_V \vec{w}_V)}{\partial t} + \nabla \cdot (\alpha\rho_V \vec{w}_V \vec{w}_V) + \nabla(\alpha p) = \qquad (2.6)$$

$$- \vec{\tau}_i \qquad \qquad \text{(interfacial friction)}$$
$$- \alpha \vec{f}_W \qquad \qquad \text{(wall friction)}$$
$$+ \psi \vec{w}_\Gamma \qquad \qquad \text{(momentum flux due to phase change)}$$
$$- \alpha\rho_V \vec{g} \qquad \qquad \text{(gravitation)}$$
$$- \alpha(1-\alpha)(\rho_L - \rho_V)\vec{g}D_h \nabla\alpha \qquad \qquad \text{(water level force)}$$
$$- \alpha(1-\alpha)\rho_m \left(\frac{\partial \vec{w}_R}{\partial t} + \nabla \vec{w}_R\right) \qquad \qquad \text{(virtual mass)}$$
$$+ S_{I,V} \qquad \qquad \text{(external momentum source terms (e.g. pumps))}$$

where:

$$\rho_m = \alpha\rho_V + (1-\alpha)\rho_L$$
$$\vec{w}_R = \vec{w}_V - \vec{w}_L$$

After spatial integration, the above six conservation equations lead to a set of first order differential equations, called the 6-equation model, or also called the 2M model. Alternatively, the separated momentum equations can be combined into an overall momentum equation for the two-phase mixture, applied within the so-called 5-equation (or 1-M) model:

$$\frac{\partial(\rho_m \vec{w}_m)}{\partial t} - \vec{w}_m \frac{\partial \rho_m}{\partial t} + \rho_m \vec{w}_m \nabla \vec{w}_m + \nabla(\alpha(1-\alpha)\frac{\rho_V \rho_L}{\rho_m}\vec{w}_R \vec{w}_R) + \nabla p = \quad (2.7)$$

$$+ \vec{f}_W \qquad\qquad\qquad\qquad\qquad \text{(wall friction)}$$
$$+ \rho_m \vec{g} \qquad\qquad\qquad\qquad\qquad \text{(gravitation)}$$
$$+ S_{I,m} \qquad\qquad\qquad \text{(external momentum source terms)}$$

where:

$$\vec{w}_m = \frac{1}{\rho_m}(\alpha \rho_V \vec{w}_V + (1-\alpha)\rho_L \vec{w}_L)$$

For the derivation of the system of differential equations in ATHLET, the following assumptions have been made:

- changes in the geometry of flow channels and structures are neglected, i.e. $\frac{dV}{dt} = 0$,
- in the energy balance equations, the potential energy contribution and the dissipation energy are neglected.

The spatial integration of the conservation equations in ATHLET is performed on the basis of a finite-volume approach.

2.1.2 System Modeling in MATLAB

In this dissertation, a one-dimensional model of the SMDFR core is built based on the equivalent parameters obtained by the coupled high-fidelity model, taking into account of the drift effect of the delayed neutron precursors by modifying the point kinetic model. The reactivity feedback, resulting from the temperature and density change in the two fluids, is considered by a linearized correlation of the introduced reactivity and the changed temperatures. After the completion the core model, a

PI controller is designed to control the power by introducing positive or negative reactivity based on the difference between the set value and the measured value of the core power.

In order to obtain basic data for the establishment of the one-dimensional model of the SMDFR core, the necessary parameters for the neutronic and thermodynamic equations are obtained from the coupled high-fidelity model. In addition, the point kinetic model has to be modified to accurately describe the process of delayed neutron precursor drift. The thermodynamic model was built under the assumption that the fuel, the pipe wall and the coolant in the core are divided into 12 parts and each part has lumped properties. The point kinetic model and the thermodynamic model were then linked by the reactivity feedback effects.

Data used for Modeling

Two types of data are required for the simulation, the neutronics data (Table 2.1) and the thermodynamics data (Table 2.2). The neutronics data are obtained by using Serpent 2.1.31 with the ENDF/B-VII nuclear data library, applying the calculated temperature and density distributions of the fuel and coolant. The thermodynamics data are calculated using the fully resolved CFD model via COMSOL Multiphysics version 5.6 [COM20c] together with its CFD module [COM20a] and heat transfer module [COM20b].

Modified Point Kinetic Model

In order to accurately capture the neutronic dynamic behavior of the SMDFR core, the point kinetic model [Kee65] with 6 delayed neutron groups (i = 1, ..., 6) has been adopted and modified. Although the fuel outside of the core is kept sub-critical, the decay process of the delayed neutron precursors has to be considered to describe its influence on the neutron distribution in the core. Traditionally, the effect of the delayed neutrons due to the flowing fuel is described by two additional terms [WLMJ18]: the precursor loss when the fuel leaves the core, and the precursor gain when the fuel re-enters the core, which are the 3rd and 4th terms on the right side of the delayed neutron precursor equations, respectively, as shown in Equation (2.8). However, the time delay term introduces some complexity into the modeling process, and some special treatments have to be applied to solve this type of equation. To eliminate this time delay term, the concentrations of the six groups of delayed neutron precursors outside of the core, $C_e i$, are defined, and then the balance of delayed neutron precursors can be described as shown in Figure 2.1. The concentrations of the six groups of delayed neutron precursors in the core are described by Equation (2.9). Since six dependent variables are introduced, six additional equations (Equation (2.10)) describing the evolution of the DNP

Table 2.1 Neutronics data for the point kinetic model

Parameters	Value
Λ (s)	1.05×10^{-6}
β_i $(-)$	$(7.91 \times 10^{-5}\,7.03 \times 10^{-4}\,5.04 \times 10^{-4}\,1.17 \times 10^{-3}\,4.57 \times 10^{-4}\,1.10 \times 10^{-4})$
β $(-)$	3.02×10^{-3}
λ $(1/s)$	$(1.27 \times 10^{-2}\,3.00 \times 10^{-2}\,1.10 \times 10^{-1}\,3.19 \times 10^{-1}\,1.18\,7.02)$
ϕ_{tot} $(1/(m^2 \cdot s))$	3.61×10^{21}
$\frac{1}{v}$ (s/m)	4.16×10^{-9}
N_0 $(1/m^3)$	1.50×10^{13}
P^i (W)	$(1.58 \times 10^{-1}\,6.42 \times 10^{-2}\,6.98 \times 10^{-2}\,7.10 \times 10^{-2}\;7.14\;\times\;10^{-2}\,7.12\;\times\;10^{-2}\,7.07\;\times\;10^{-2}\,7.00 \times 10^{-2}\,6.91 \times 10^{-2}\,6.76 \times 10^{-2}\,6.49 \times 10^{-2}\,1.52 \times 10^{-1})$
P_0 (W)	1.00×10^8
τ_c (s)	5.43
τ_e (s)	10.0
λ_c $(1/s)$	0.184
λ_e $(1/s)$	0.1
α_f $(1/K)$	-2.08×10^{-4}
α_c $(1/K)$	-8.28×10^{-6}
ρ_0 $(\$)$	1.11×10^{-3}
$C_{i,0}$ $(1/m^3)$	$(4.29 \times 10^{14}\,4.15 \times 10^{15}\,3.83 \times 10^{15}\,1.16 \times 10^{16}\,5.70 \times 10^{15}\,1.53 \times 10^{15})$
$Ce_{i,0}$ $(1/m^3)$	$(7.01 \times 10^{14}\,5.88 \times 10^{15}\,3.36 \times 10^{15}\,5.09 \times 10^{15}\,8.20 \times 10^{14}\,3.96 \times 10^{13})$

concentrations outside of the core, together with the neutron density equation (Equation (2.11)), have to be added to close the set of equations for the point kinetic model.

$$\frac{dC_i(t)}{dt} = \frac{\beta_i}{\Lambda} \cdot N(t) - \lambda_i C_i(t) - \lambda_c C_i(t) + \lambda_c C_i(t - \tau_e) \cdot e^{-\lambda_i \tau_e} \tag{2.8}$$

$$\frac{dC_i(t)}{dt} = \frac{\beta_i}{\Lambda} \cdot N(t) - \lambda_i C_i(t) - \lambda_c C_i(t) + \lambda_e Ce_i(t) \tag{2.9}$$

$$\frac{dCe_i(t)}{dt} = -\lambda_i Ce_i(t) - \lambda_e Ce_i(t) + \lambda_c C_i(t) \tag{2.10}$$

$$\frac{dN(t)}{dt} = \frac{(\rho(t) - \beta)}{\Lambda} \cdot N(t) + \sum_{i=1}^{6} \lambda_i C_i(t) \tag{2.11}$$

Table 2.2 Thermodynamics data

Parameters	Distribution zone	Core Zone (1 node)	Collection zone
Fuel pipes (#)	–	–	1027
Coolant pipes (#)	2166	2166	–
A_{fw} (m^2)	13.6	9.0	13.6
A_{wc} (m^2)	10.9	10.3	10.9
$\dot{m}_f$ (kg/s)	327.7	327.7	327.7
$\dot{m}_c$ (kg/s)	5550.5	5550.5	5550.5
M_f (kg)	307.7	116.4	307.7
M_w (kg)	39.3	31.1	39.3
M_c (kg)	219.9	770.5	219.9
$c_{p,f}$ $(J/(kg \cdot K))$	400	400	400
$c_{p,w}$ $(J/(kg \cdot K))$	690	690	690
$c_{p,c}$ $(J/(kg \cdot K))$	140.2	140.2	140.2
Heat taken by coolant (J)	18,744,000	5,970,000	16,956,000
h_{fw} $(W/(K \cdot m^2))$	6546.0	4305.1	7629.5
h_{wc} $(W/(K \cdot m^2))$	27,405.4	13,833.4	32,177.3
T_f^{in} (K)	1300.0	–	–
T_c^{in} (K)	973.0	–	–

The initial value of ρ, C_i and Ce_i can be calculated by solving the governing equations Equations (2.9)–(2.11), applying a steady state condition by setting the time derivative terms equal to zero, as shown in Equations (2.12)–(2.14).

$$\rho_0 = \beta - \sum_{i=1}^{6} \frac{\beta_i \cdot (\lambda_i + \lambda_e)}{\lambda_i + \lambda_e + \lambda_c} \tag{2.12}$$

$$C_{i,0} = \frac{\beta_i}{\Lambda} \cdot N_0 \cdot \frac{\lambda_i + \lambda_e}{(\lambda_i + \lambda_e + \lambda_c) \cdot \lambda_i} \tag{2.13}$$

$$Ce_{i,0} = C_{i,0} \cdot \frac{\lambda_c}{\lambda_i + \lambda_e} \tag{2.14}$$

Thermodynamic Model

Considering the geometric structure of the core, 12 nodes are defined: Node 1 for the distribution zone, Node 2–11 for the core zone (from bottom to top), and Node 12

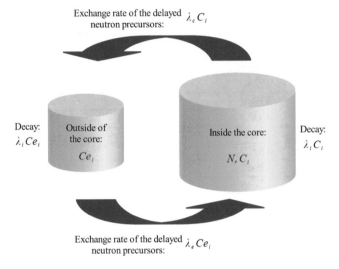

Exchange rate of the delayed $\lambda_e C_i$
neutron precursors:

Decay: Outside of   Inside the core: Decay:
$\lambda_i Ce_i$ the core: $\lambda_i C_i$
                Ce_i N, C_i

Exchange rate of the delayed $\lambda_e Ce_i$
neutron precursors:

Figure 2.1 Schematic of the balance of delayed neutron precursors

for the collection zone, as shown in Figure 2.2. Each node has a height of 0.2 m. The energy balance of Node 1 is described by Equations (2.15)–(2.17). For the remaining nodes (Node 2–12, i = 2, ..., 12), the heat transfer process is described by Equations (2.18)–(2.20). As shown in Figure 2.2, the heat generated by the nuclear fission is transferred from the fuel to the pipe wall and then to the primary coolant. Finally, the fission energy is absorbed by the primary coolant and transferred to the secondary coolant for utilization. The power is assumed to be proportional to the neutron density, as shown in Equation (2.21).

$$M_f^1 \cdot c_{p,f}^1 \cdot \frac{dT_f^1(t)}{dt} = P^1(t) + \dot{m}_f \cdot c_{p,f}^1 \cdot (T_f^{in} - T_f^e(1)) - h_{fw}^1 \cdot A_{fw}^1 \cdot (T_f^1 - T_w^1)$$

$$(2.15)$$

$$M_w^1 \cdot c_{p,w}^1 \cdot \frac{dT_w^1(t)}{dt} = h_{fw}^1 \cdot A_{fw}^1 \cdot (T_f^1 - T_w^1) - h_{wc}^1 \cdot A_{wc}^1 \cdot (T_w^1 - T_c^1) \qquad (2.16)$$

$$M_c^1 \cdot c_{p,c}^1 \cdot \frac{dT_c^1(t)}{dt} = \dot{m}_c \cdot c_{p,c}^1 \cdot (T_c^{in} - T_c^e(1)) + h_{wc}^1 \cdot A_{wc}^1 \cdot (T_w^1 - T_c^1) \qquad (2.17)$$

$$M_f^i \cdot c_{p,f}^i \cdot \frac{dT_f^i(t)}{dt} = P^i(t) + \dot{m}_f \cdot c_{p,f}^i \cdot (T_f^e(i-1) - T_f^e(i)) - h_{fw}^i \cdot A_{fw}^i \cdot (T_f^i - T_w^i)$$

$$(2.18)$$

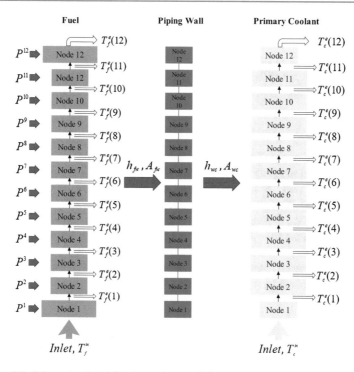

Figure 2.2 Schematic of nodalization and energy balance

$$M_w^i \cdot c_{p,w}^i \cdot \frac{dT_w^i(t)}{dt} = h_{fw}^i \cdot A_{fw}^i \cdot (T_f^i - T_w^i) - h_{wc}^i \cdot A_{wc}^i \cdot (T_w^i - T_c^i) \tag{2.19}$$

$$M_c^i \cdot c_{p,c}^i \cdot \frac{dT_c^i(t)}{dt} = \dot{m}_c \cdot c_{p,c}^i \cdot (T_c^e(i-1) - T_c^e(i)) + h_{wc}^i \cdot A_{wc}^i \cdot (T_w^i - T_c^i) \tag{2.20}$$

$$P(t) = P_0 \cdot \frac{N(t)}{N_0} \tag{2.21}$$

Reactivity Feedback

The change in fuel or coolant temperature introduces additional reactivity, called the reactivity feedback effect, which has an important effect on the operation of the reactor. The reactivity $\rho(t)$ consists of the initial reactivity, the reactivity introduced by the variation of the fuel and coolant temperatures, and the externally introduced reactivity from control rods or other sources, as shown by Equation (2.22):

$$\rho(t) = \rho_0 + \alpha_f(\overline{T}_f(t) - \overline{T}_{f,0}) + \alpha_c(\overline{T}_c(t) - \overline{T}_{c,0}) + \rho_{insert} \qquad (2.22)$$

where α_f is the temperature feedback coefficient of fuel; α_c is the temperature feedback coefficient of the coolant; $\overline{T}_{f,0}$ and $\overline{T}_{c,0}$ are the initial mean temperatures of the fuel and coolant, respectively; and $\overline{T}_f(t)$ and $\overline{T}_c(t)$ denote the real time mean temperatures.

PI Controller

In this dissertation, the PI controller is adopted for the control system. A variation of the proportional integral derivative (PID) controller, the proportional integral (PI) controller uses only the proportional and integral terms. The difference between the set value and the real (measured) value, $e(t)$ (Equation (2.23)), is given to the PI controller as the feedback error, and then the value of the controller output $u(t)$ is calculated in the time domain from the feedback error by Equation (2.24) and then fed into the system as the manipulated variable input. It is clear that the two parameters K_p and K_i have a significant influence on the system response and, thus, should be optimized for a better control performance.

$$e(t) = P_{set} - P_{real} \qquad (2.23)$$

$$u(t) = K_p e(t) + K_i \int_0^t e(t)dt \qquad (2.24)$$

2.1.3 Neutron Transport Modeling in the Monte-Carlo Code Serpent

In the explicit coupling scheme, the neutron transport is solved by the continuous-energy Monte-Carlo reactor physics code: Serpent, which has been developed at VTT Technical Research Centre of Finland since 2004, and is currently used in over 100 universities and research organizations worldwide [LPV+15].

The basic geometry routine in Serpent is based on a three-dimensional constructive solid geometry model, which is built from elementary quadratic and derived surface types to form two- or three-dimensional cells. The geometry can be partitioned into multiple levels using universes, transformations and repeating structures. The neutron interaction data are read by Serpent from cross section libraries in continuous-energy ACE format, which form the basis for the physical laws in the transport simulation. [LPV+15]

Since Serpent was employed for the group constant generation and power distribution, only the criticality calculation was performed, and its feature of burnup calculation feature is not relevant and outside the scope of the dissertation.

2.1.4 Neutron Transport Modeling in COMSOL Multiphysics

In the implicit coupling scheme, the neutron transport is calculated by the multigroup neutron diffusion equations in COMSOL Multiphysics:

$$\frac{1}{v^g}\frac{\partial \phi^g(r,t)}{\partial t} = \tag{2.25}$$

$$\nabla \cdot (D^g(r,t) \nabla \phi^g(r,t)) \qquad \text{(neutron diffusion)}$$

$$- \Sigma_{a,g}(r,t)\phi^g(r,t) \qquad \text{(neutron absorption)}$$

$$+ \sum_{g'=1}^{G} \Sigma_s^{g' \to g}(r,t)\phi^{g'}(r,t) \qquad \text{(neutron scattering)}$$

$$+ \chi_p^g \sum_{g'=1}^{G} (1 - \beta^{g'})\nu \Sigma_f^{g'}(r,t))\phi^{g'}(r,t) \qquad \text{(prompt neutrons)}$$

$$+ \sum_{j=1}^{J} \chi_{d,j}^g \lambda_j C_j(r,t), \qquad \text{(delay neutrons)}$$

$$\forall g \in [1,G], \qquad \text{(G groups)}$$

where G is the total number of energy groups and J is the total number of delayed neutron precursor groups.

In the SMDFR, the fuel is fluid and follows through the core, which means that the Delayed Neutron Precursors (DNPs) are not confined inside the core, but are continuously exchanged with the contents outside of the core during operation. Due to this drift effect, the delayed neutrons are significantly influenced by the variation of the DNPs. To correctly describe the neutron transport, the concentrations of the DNPs have to be explicitly solved by the following equations:

$$\frac{\partial C_j(r,t)}{\partial t} = \nabla \cdot (D_j(r,t) \nabla C_j(r,t)) - u \cdot \nabla C_j(r,t) + \sum_{g'=1}^{G} \beta_j^{g'} \nu \Sigma_f^{g'}(r,t)) \phi^{g'}(r,t) - \lambda_j C_j(r,t),$$

(2.26)

$$\forall j \in [1, J], \qquad \text{(J delayed neutron precursor groups)}$$

where G is the total number of energy groups and J is the total number of delayed neutron precursor groups.

2.1.5   CFD Modeling in Ansys CFX and COMSOL Multiphysics

The heat transfer between two fluids separated by a solid wall is the main thermal process to be simulated, with both fluids the molten salt and the liquid lead having high Reynolds numbers (molten salt: 1.17×10^4, liquid lead: 1.28×10^4). The governing equations for both COMSOL Multiphysics and Ansys CFX are the same and will be introduced using COMSOL Multiphysics as an example. In COMSOL Multiphysics, the physical COMSOL Multiphysics interface "Conjugate Heat Transfer: Turbulent Flow, k-ϵ" was selected for the thermal-hydraulics calculation, which combines the "Heat Transfer in Solids and Fluids" interface and the "Turbulent Flow, k-ϵ" interface and, thus, can be used to simulate the coupling between heat transfer and turbulent flow.

Three compressibility options are available in COMSOL Multiphysics: compressible flow (Ma < 0.3), weakly compressible flow, and incompressible flow. The compressible flow option makes no assumptions about the system and takes into account any dependence of the fluid properties on the variables. The equations for weakly compressible flow are the same as those of compressible flow, except that the density is evaluated at the reference absolute pressure. For the incompressible flow option, the density is considered constant and evaluated at the reference temperature and pressure. [Car16] Since the density of molten salt is strongly dependent on its temperature, the weakly compressible Reynolds-Averaged Navier-Stokes (RANS) equations were adopted as a compromise between computational cost and accuracy:

$$\rho(\mathbf{u} \cdot \nabla)\mathbf{u} = \nabla \cdot [-p\mathbf{I} + \mathbf{K}] + \mathbf{F} \tag{2.27}$$

$$\nabla \cdot (\rho\mathbf{u}) = 0 \tag{2.28}$$

$$\mathbf{K} = (\mu + \mu_T)(\nabla\mathbf{u} + (\nabla\mathbf{u})^T) - \frac{2}{3}(\mu + \mu_T)(\nabla \cdot \mathbf{u}) - \frac{2}{3}\rho k\mathbf{I} \tag{2.29}$$

where $\mathbf{K}$ is the viscosity term taking into account the interactions between the fluctuating parts of the velocity field. In order to close the above equations, two additional transport equations with two dependent variables, the turbulent kinetic energy k and the turbulent dissipation rate ϵ, are introduced:

$$\rho(\mathbf{u} \cdot \nabla)k = \nabla \cdot [(\mu + \frac{\mu_T}{\sigma_k}) \nabla k] + P_k - \rho\epsilon \tag{2.30}$$

$$\rho(\mathbf{u} \cdot \nabla)\epsilon = \nabla \cdot [(\mu + \frac{\mu_T}{\sigma_\epsilon}) \nabla \epsilon] + C_{\epsilon 1}\frac{\epsilon}{k} P_k - C_{\epsilon 2}\rho\frac{\epsilon^2}{k} \tag{2.31}$$

where μ_T and P_k are the turbulent viscosity and the production term and are given by:

$$\mu_t = \rho C_\mu \frac{k^2}{\epsilon} \tag{2.32}$$

$$P_k = \mu_T[\nabla\mathbf{u} : (\nabla\mathbf{u} + (\nabla\mathbf{u})^T) - \frac{2}{3}(\nabla \cdot \mathbf{u})^2] - \frac{2}{3}\rho k \nabla \cdot \mathbf{u} \tag{2.33}$$

The model's constants used are: $C_{\epsilon 1} = 1.44$, $C_{\epsilon 2} = 1.92$, $C_\mu = 0.09$, $\sigma_k = 1.0$, $\sigma_\epsilon = 1.3$.

For the heat transfer in solids and fluids, the following equations are adopted:

$$\rho C_p \mathbf{u} \cdot \nabla T + \nabla \cdot \mathbf{q} = Q \tag{2.34}$$

$$\mathbf{q} = -k \nabla T \tag{2.35}$$

and the heat flux between the fluid with temperature T_f and the wall with temperature T_w is calculated by:

$$-\mathbf{n} \cdot \mathbf{q} = \rho C_p u_\tau \frac{T_w - T_f}{T^+} \tag{2.36}$$

Since there is no recommended turbulent Prandtl number model for liquid lead flow, the Kays–Crawford model is adopted:

$$Pr_T = \left\{\frac{1}{2Pr_T\infty} + \frac{0.3}{\sqrt{Pr_T\infty}}\frac{C_p\mu_T}{k} - (0.3\frac{C_p\mu_T}{k})^2(1 - e^{-k/(0.3C_p\mu_T\sqrt{Pr_T\infty})})\right\}^{-1} \tag{2.37}$$

where the Prandtl number at infinity is $Pr_T\infty = 0.85$.

2.2   Multiscale Coupling Method Adopted in SESAME Project

The coupled model of the TALL-3D facility for ATHLET-ANSYS CFX was generated by GRS [PGL+19] and is shown in Figure 2.3. The coupling approach [PLW+11] is based on a domain decomposition strategy. The whole computational domain is divided into subdomains and each subdomain is simulated by either ATHLET or ANSYS CFX. Since the detailed thermal-hydraulic behaviors in the 3D test section need to be investigated, it was modeled by ANSYS CFX (Figure 2.3, right) to provide an adequate simulation of the multidimensional phenomena in this particular subdomain. An 1° rotational symmetry sector of the TS was represented here. Previous analyses showed that the consideration of the solid test section structures in the CFD model is of great importance for the results [Pap16]. Therefore, the metal plate, the solid walls, the heater and even the insulation were explicitly modeled with ANSYS CFX. The resulting mesh was systematically refined and mesh sensitivity studies were performed according to the OECD Best Practice Guidelines [MCS+07]. The mesh selected for the final 1o simulations had 109000 elements (Figure 2.3, right). The ATHLET-ANSYS CFX calculation presented here was performed with the SST turbulence model [Men94]. At the inlet of the test section a turbulence intensity of 5% was specified.

Other subdomains were modeled by ATHLET (Figure 2.3, left), since the macroscopic thermal-hydraulic descriptions of these regions are sufficient to represent their influence on the system behavior. Four priority chains (flow paths) were created for the simulation of the experimental facility with ATHLET. The pressure was controlled by a time-dependent volume simulating the function of the TALL-3D expansion tank. The secondary circuit was represented in a simplified way with three pipes as a heat sink. Mass flow rate and enthalpy were specified at the inlet pipe using a "FILL" component, and the circuit ended with a time-dependent volume. The entire facility model consisted of approximately 170 control volumes (Figure 2.3, left). For the coupled ATHLET-ANSYS CFX simulations an adaptive time stepping scheme (time step sizes between 0.05 s and 0.1 s) was chosen to ensure numerical convergence and stability.

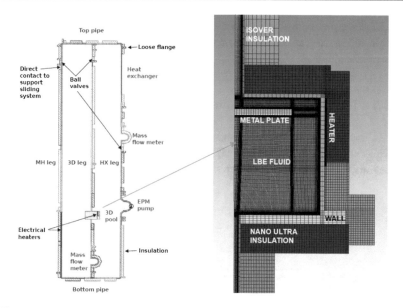

Figure 2.3 TALL-3D coupled ATHLET (left) and ANSYS CFX (right) model

2.3 Multiphysics Coupling Approaches Adopted in the Analysis of SMDFR

Since the nuclear chain reaction and the heat transfer take place simultaneously, a strong coupling effect is expected and has to be considered by using appropriate coupling approaches.

First, a single flow channel was selected for the analysis of its steady-state and transient local behaviors, based on its numerical simulation by means of the coupled 3D neutronics/thermal-hydraulics time-dependent model. A fully implicit coupling scheme was developed and applied in COMSOL Multiphysics® [COM20c], which is described in detail in Section 2.3.1.

Once the local behaviors were well understood, the full core model of the SMDFR was built to study the global behaviors of the reactor core. Since the geometry was quite complex, a high-fidelity explicit coupling scheme (Section 2.3.2) was chosen: the neutronics field was solved by the continuous-energy Monte-Carlo reactor physics code: Serpent, and the thermal-hydraulic field was solved by the CFD code: COMSOL Multiphysics. It should be noted that in Serpent the neutron transport was solved by the continuous-energy cross section and in COMSOL Multiphysics the

RANS equations were solved fully for the whole domain without any geometrical simplification. As a result, few assumptions are made and the delivered results can be considered to have a high fidelity.

2.3.1 Implicit Coupling Scheme

In the implicit coupling scheme, the neutron transport is modeled by the multi-group neutron diffusion equations in COMSOL Multiphysics. To solve the diffusion equations, a Serpent model with the same geometry and material composition has

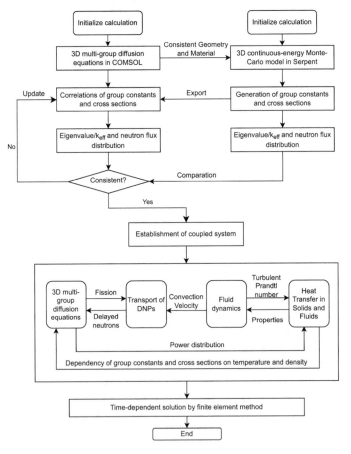

Figure 2.4 Principle of implicit coupling scheme

to be created prior to the coupling to generate the necessary correlations of group constants and cross sections. After solving the diffusion equations, the obtained eigenvalue/k_{eff} and neutron flux/power distribution have to be checked against those of the Serpent model. If they are not consistent, the correlations of group constants and cross sections have to be updated and the eigenvalue/k_{eff} and neutron flux/power distribution are recalculated. When consistency is achieved, other physical fields are added to form the coupled system. In the single channel model there are four physical processes to be considered: the neutron transport, the transport of the DNPs, the fluid dynamics, and the heat transfer in solids and fluids. As shown in Figure 2.4, mutual interactions between different fields have to be considered. The behavior of neutron transport affects the production of DNPs, while the decay of the DNPs affects the production of delayed neutrons. The convection of DNPs is determined by the fuel velocity, which is given by the results of the fluid dynamics calculation. The fluid dynamics and the heat transfer fields are linked by the turbulent Prandtl number and the fluid properties. The power distribution is given by the solution of the diffusion equations, while the group constants and cross sections in the diffusion equations depend on the fluid temperature and density, which are determined by the heat transfer process.

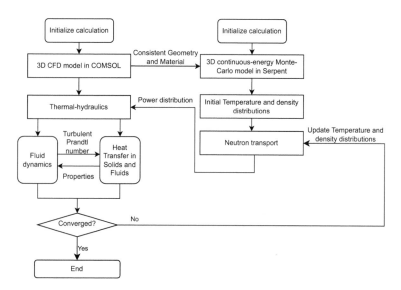

Figure 2.5 Principle of explicit coupling scheme

2.3.2   Explicit Coupling Scheme

In the explicit coupling scheme (Figure 2.5), the neutron transport is solved by the 3D continuous-energy Monte-Carlo model in Serpent and the thermal-dynamics is solved by the 3D CFD model in COMSOL, and both models have a consistent geometry and material composition. The initial temperature and density distributions are given to the Serpent model to start the neutron transport calculation, which provides a power distribution as a heat source for the CFD model. The CFD model is then solved by coupling the fluid dynamics and the heat transfer in solids and fluids. The results obtained are compared with those of the previous iteration. If they do not converge, the temperature and density distributions for the Serpent model are updated, resulting in an updated power distribution. A new simulation of

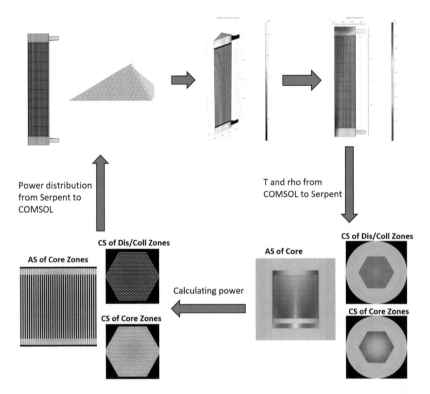

Figure 2.6 Schematic of explicit coupling scheme

the CFD model is performed with the updated power distribution. The simulation is terminated when the convergence criteria are met, i.e. the temperature and density distributions are no longer changed. The schematic of this explicit coupling scheme, which is implemented in the SMDFR full core model, is shown in Figure 2.6.

2.4 Uncertainty Quantification and Sensitivity Analysis Methods

The different approaches of uncertainty quantification and sensitivity analysis are summarized in Section 2.4.1 and Section 2.4.2. And the sampling techniques and surrogate mode are introduced in Section 2.4.3 and Section 2.4.4. Finally, the chains of calculations for forward and inverse uncertainty propagation are described in Section 2.4.5.

2.4.1 Uncertainty Quantification

Uncertainty quantification aims at investigating the variability of the output quantities of interest by taking into account the uncertainties encountered throughout the modeling process. There are two main types of uncertainty quantification problems: forward propagation of uncertainty (where the various sources of uncertainty are propagated through the model to predict the overall uncertainty in the system response) and the inverse assessment of model uncertainty and parameter uncertainty (where the model parameters are simultaneously calibrated using test data).

Forward uncertainty propagation Forward uncertainty propagation is the quantification of uncertainties in system outputs propagated from uncertain inputs. In this dissertation, the primary goal of uncertainty propagation is to evaluate the full probability distribution of the simulation outputs. In this particular scenario, two uncertainty analysis methods are presented: Monte-Carlo Method and Tolerance Limits. The Monte-Carlo method can be applied if the code or surrogate model runs fast enough, given the available computing capacity, that thousands of runs can be performed within an acceptable time. Otherwise, the tolerance limits should be used to reduce the number of runs to less than one hundred.

Inverse uncertainty propagation Inverse uncertainty propagation aims to estimate the discrepancy between the experiment and the mathematical model (bias correction) and to estimate the values of unknown parameters (calibration), which is implemented during the quantification of the MATLAB model in Section 6.2.

Monte-Carlo Method

Statistical methods can be used to compute and return the statistical quantile with the associated confidence interval. The estimation is performed by generating a sample of limited size following the distributions of the input parameters, performing the appropriate calculations/simulations and processing the output data with appropriate statistical techniques. This method of uncertainty analysis is called the Monte-Carlo method. According to the central limit theorem, the accuracy of the estimate depends on the sample size, and the sample size usually ranges from a few hundred to a few thousand. For the numerical simulation of TG03.S301.03 with a coupled code system (ATHLET-CFX), the typical computation time for a was was about 10 hours using 8 Central Processing Units (CPUs). Since the computational capacity and the available licenses of the codes are quite limited compared to the total computational demand, the direct uncertainty quantification using the Monte-Carlo method is obviously not feasible. In this case, a surrogate mode of the original computer mode should be built based on the results of the performed runs, and the computational demand would be dramatically reduced by using the surrogate model.

As a substitution of the statistical quantile and the associated confidence interval, the tolerance limits can be used to save computational resources. The number of runs required depends on the population coverage, the confidence level and the order. For two-sided, it starts at 93 for $(0.95, 0.95)$, and the sample size becomes much larger for $(0.99, 0.99)$, as shown in Table 2.3.

The sample mean can be calculated by:

$$\widehat{E(Y_j)} = \bar{y}_j = \frac{1}{N} \sum_{i=1}^{N} y_{j,i} \tag{2.38}$$

where j is the index of the j-th output quantity, N is the sample size; the hat of $\widehat{E(Y_j)}$ indicates that the mean values is estimated based on a sample of observed values. This estimator differs from the true value $E(Y_j)$, which is linked to the original population, associated to the random variable Y_j.

The sample variance can be calculated by:

$$\widehat{Var(Y_j)} = \frac{1}{N-1}\sum_{i=1}^{N}(y_{j,i} - \bar{y}_j)^2 \tag{2.39}$$

this estimator is unbiased. The sample standard deviation is

$$\hat{\sigma}_j = \sqrt{\hat{\sigma}_j^2} = \sqrt{\widehat{Var(Y_j)}} \tag{2.40}$$

The approximate confidence interval for the mean can be calculated by:

$$CI_{1-\alpha} = \widehat{E(Y_j)} \pm z_{1-\frac{\alpha}{2}} \frac{\hat{\sigma}_j}{\sqrt{N}} \tag{2.41}$$

where $1-\alpha$ is the confidence level and $z_{1-\frac{\alpha}{2}}$ is the $1-\frac{\alpha}{2}$ quantile of a standard normal random variable. Typically α is set equal to 0.01 (99% confidence) or 0.05 (95% confidence). Confidence intervals indicate that the true value of the estimated quantity lies between the lower and upper bounds for a given level of confidence. This formula is an approximation and is considered valid when $N > 30$.

The approximate confidence interval for a quantile is based on order statistics. The integers r and t that satisfy the following equation have to be estimated:

$$P\left(y(r) \leq Q_p \leq y(t)\right) \geq 1-\alpha \tag{2.42}$$

with Q_p the p-th quantile and $1-\alpha$ the confidence level. The values of r and t can be approximated as follows:

$$\check{r} = Np + z_{\frac{\alpha}{2}}\sqrt{Np(1-p)} \tag{2.43}$$

$$\check{t} = Np + z_{1-\frac{\alpha}{2}}\sqrt{Np(1-p)} \tag{2.44}$$

The obtained quantities $\check{r}$ and $\check{t}$ are most likely not integers. The value of r and t is obtained by rounding $\check{r}$ and $\check{t}$ up to the next higher integer.

Tolerance Limits

Instead of using the statistical quantile with a confidence interval based on the Monte-Carlo method, which is quite computationally expensive, Glaeser [Gla08] proposed tolerance limits to quantify the uncertainty of computer code results with only about a hundred runs. The concept of tolerance limits, also known as a tolerance interval, is a statistical interval within which a specified proportion of a sampled population falls at a given confidence level. As defined by Krishnamoorthy and Mathew [KM09], there are two types of tolerance limits: one-sided tolerance limit and two-sided tolerance limits.

- **One-sided tolerance limits:** A $(p, 1 - \alpha)$ upper/lower tolerance limit is simply an $1 - \alpha$ upper/lower confidence limit for the $(100 * p)^{th}$ percentile of the population, meaning that at least $100 * p$ percent of the population would fall below/exceed the upper/lower tolerance limit with a probability of $1 - \alpha$.
- **Two-sided tolerance limits:** A $(p, 1 - \alpha)$ two-sided tolerance limits are simply the $1 - \alpha$ two-sided confidence limits for the $(100 * p)^{th}$ percentile of the population, meaning that at least $100 * p$ percent of the population would fall within the two-sided tolerance limits with a probability of $1 - \alpha$.

A formulation of the confidence level α is proposed by Noether [Noe67]:

$$\alpha \leq \sum_{k=0}^{r+s-1} \binom{N}{k} (1 - p)^k p^{N-k} \tag{2.45}$$

For one-sided upper/lower tolerance limit, r/s is set to 0. For two-sided tolerance limits, r should normally be equal to s. In this dissertation, first-order two-sided tolerance limits are chosen for uncertainty quantification, so $r = s = 1$. Given r, s, p (population proportion of interest) and α, the value of N (sample size) is limited by this formula and the number of required runs should be greater than the smallest value of N.

The values of the required sample size N based on $1 - \alpha = 0.90 \rightarrow 0.99$ and $p = 0.90 \rightarrow 0.99$ are summarized in Table 2.3.

For all the following uncertainty quantification studies, including the blind and open phase of the test case TG03.S301.03 and the SMDFR models, the sample size was set to 93. All the runs for uncertainty quantification were successful, so the value of the confidence level and the population proportion of interest p are set to be 0.95 and 0.95, respectively.

Table 2.3 Sample sizes for tolerance limits ($r = s = 1$)

N_{min} \ $1 - \alpha$ p	0.90	0.91	0.92	0.93	0.94	0.95	0.96	0.97	0.98	0.99
0.90	38	39	41	42	44	46	49	52	56	64
0.91	42	44	45	47	49	51	54	58	63	71
0.92	48	49	51	53	55	58	61	65	71	81
0.93	55	56	58	61	63	66	70	75	81	92
0.94	64	66	68	71	74	78	82	88	95	108
0.95	77	79	82	85	89	93	99	105	115	130
0.96	96	100	103	107	112	117	124	132	144	164
0.97	129	133	138	143	149	157	166	177	193	219
0.98	194	200	207	215	225	236	249	266	290	330
0.99	388	401	416	432	451	473	500	534	581	662

2.4.2   Sensitivity Analysis

The purpose of sensitivity analysis is to study how the output variability of a system that can be attributed to the uncertainty in the input. There are four main types of sensitivity analysis methods: graphical methods, screening analysis, regression-based analysis and variance-based analysis.

Graphical Methods

Graphical methods, such as scatter plots and cobweb plots, are very easy to implement and can be applied regardless of the sampling techniques. They provide a quick and easy way to get a first opinion of the sensitivity analysis. However, since these types of methods are not able to provide a quantitative sensitivity assessment and rely too much on subjective impression, they are not suitable for the sensitivity analysis in this dissertation.

Screening Analysis

The method of elementary effects proposed by Morris [Mor91] is widely used for screening analysis: a process that identifies the most important variables in a sample by a certain manipulation of the sample. It relies on the One at A Time (OAT) sampling technique to assess the sensitivity of the input variables. Because of the use of the OAT sampling technique, the samples generated for the screening analysis are not suitable for the calculating tolerance limits or the training of a surrogate model. Although a qualitative assessment of the level of significance can be made using the screening analysis, it is not suitable for the quantitative sensitivity assessment compared to the regression-based analysis and variance-based analysis. Considering the high computational cost, this technique is not preferred for sensitivity analysis of a computational model using 1D-3D coupled codes.

Regression-based Analysis

Regression-based techniques are well suited for the quantitative sensitivity assessment when the system is linear and have a very low computational cost compared to other methods. When the sample is generated using a random sampling technique, the simulation results for regression-based techniques can be used for both the calculation of tolerance limits and the training of a surrogate model, which is quite preferred for the cases where the computational cost is very high (exactly the case of coupled 1D-3D codes).

In this dissertation, the Standard Regression Coefficient (SRC), which can be estimated by calculating the Person's Ordinary Correlation Coefficients (Formula (2.46)) based on the standardized variables (Formula (2.47)), is chosen as the correlation measure. The squared value of the SRC is equal to the fraction of the output variance that is linearly explained by the input.

$$r(X_j, Y) = \frac{\sum_{i=1}^{N}(y_i - \bar{Y})(x_{i,j} - \bar{X}_j)}{\left[\sum_{i=1}^{N}(x_{i,j} - \bar{X}_j)^2 \sum_{i=1}^{N}(y_i - \bar{Y})^2\right]^{\frac{1}{2}}} \qquad (2.46)$$

with:

- r is the standard regression coefficient
- $x_{i,j}$ is the j-th input variable
- y_i is the output variable
- $\bar{X}_j$ is the average of the jth input variable
- $\bar{Y}$ is the average of the output variable

$$\tilde{Z} = \frac{Z - E(Z)}{\sigma(Z)} \qquad (2.47)$$

with:

- σ is the standard deviation
- E is expected value

However, the coefficient of multiple determination (Formula (2.48)) [GKHL+08]

$$R^2 = (r(Y, X_1), ..., r(Y, X_k))[Corr(X)]^{-1}(r(Y, X_1), ..., r(Y, X_k))^T$$

$$(2.48)$$

with:

- $Corr(X)$ is the correlation matrix of the input variables
- $r(Y, X_j)$ is the correlation coefficient between Y and the j-th input variable

has to be greater than 0.6 to ensure that a significant portion of the output variability can be explained by the input variability, assuming a linear relationship. In cases where the coefficient of multiple determination is less than 0.6, regression-based analysis is no longer appropriate and the nonlinear relationship has to be accounted for by variance-based analysis.

Variance-Based Analysis

The variance-based method (Sobol method), which is originally proposed by Ilya M. Sobol, is based on the decomposition of the output variance of the model or system. Certain percentages reflecting the contribution of each input variable to the output variance can be directly interpreted as measures of sensitivity and provide an accurate quantification of global sensitivity. Furthermore, this method is well suited for the quantitative sensitivity assessment regardless of whether the system is linear or nonlinear and can handle the interaction between input variables. However the computational cost is quite high: $(k+2)*N$, where k is the number of input variables and N is the sample size, which is typically greater than 500. Thus, such techniques cannot be performed directly on the coupled code due to the high computational cost. By building a fast running surrogate model, the implementation of variance-based analysis becomes possible and is more preferred over other techniques.

Treating the model as a black box, it can be described by a function $Y = f(X)$, where X is a vector of N model inputs (sample) $\{X_1, X_2, ..., X_N\}$ and Y is the N corresponding model outputs. It is further assumed that the inputs are independent and uniformly distributed within the unit hypercube, and that $f(X)$ is decomposable:

$$Y = f_0 + \sum_{i=1}^{N} f_i(X_i) + \sum_{i<j}^{N} f_{ij}(X_i, X_j) + \cdots + f_{1,2,...,N}(X_1, X_2, ..., X_N)$$

$$(2.49)$$

with:

- f_0 is a constant
- f_i is a function of X_i
- f_{ij} is a function of X_i and X_j, etc.

A condition of this decomposition is:

$$\int_0^1 f_{i_1,i_2,...,i_s}(X_{i_1}, X_{i_2}, ..., X_{i_s})dX_k = 0, \, for \, k = i_1, i_2, ..., i_s \tag{2.50}$$

which means all the term in the functional decomposition are orthogonal. The terms of the functional decomposition can be defined in terms of conditional expected values:

$$f_0 = E(Y) \tag{2.51}$$

$$f_i(X_i) = E(Y \mid X_i) - f_0 \tag{2.52}$$

$$f_{ij}(X_i, X_j) = E(Y \mid X_i, X_j) - f_0 - f_i(X_i) - f_j(X_j) \tag{2.53}$$

where it is implied that $f_i(X_i)$ is the effect of varying X_i alone (main effect of X_i) and $f_{ij}(X_i, X_j)$ is the effect of varying X_i and X_j simultaneously and individual, which is known as a second-order interaction. Higher-order terms have analogous characteristics. Assuming that the $f(X)$ is square-integrable, the functional decomposition can be squared and integrated to give:

$$\int_0^1 f^2(X)dX - f_0^2 = \sum_{s=1}^{N} \sum_{i_1 < \cdots < i_s}^{N} \int f_{i_1 \cdots i_s}^2 dX_{i_1} \cdots dX_{i_s} \tag{2.54}$$

where the left side is equal to the variance of Y, and the terms on the right side are variance terms, now decomposed with respect to sets of X_i. Finally the decomposition of the variance expression is given:

$$Var(Y) = \sum_{i=1}^{N} V_i + \sum_{i<j}^{N} V_{ij} + \cdots + V_{12\ldots N} \tag{2.55}$$

with

$$V_i = Var_{X_i}(E_{X_{\sim i}}(Y \mid X_i)) \tag{2.56}$$

$$V_{ij} = Var_{X_{ij}}(E_{X_{\sim ij}}(Y \mid X_i, X_j)) - V_i - V_j \tag{2.57}$$

and so on.

The notation $X_{\sim i}$ indicates the set of all variables except X_i. This variance decomposition shows that the variance of the model output can be decomposed into terms attributable to each input variable, as well as the interaction between them, and all terms sum to the total variance of the model output.

The main effect index (first-order sensitivity index) is stated as:

$$S_i = \frac{V_i}{Var(Y)} \tag{2.58}$$

implying that:

$$\sum_{i=1}^{N} S_i + \sum_{i<j}^{N} S_{ij} + \cdots + S_{12\ldots N} = 1 \tag{2.59}$$

The total effect index, which measures the contribution of X_i to the output variance, including all variance caused by its interactions, of any order, with any other input variables, is given as:

$$S_{Ti} = \frac{E_{X_{\sim i}}(Var_{X_i}(Y \mid X_{\sim i}))}{Var(Y)} \tag{2.60}$$

implying that:

$$\sum_{i=1}^{N} S_{Ti} \geq 1 \tag{2.61}$$

Using the Sobol sequence, an $N \times 2d$ sample matrix can be generated with respect to the probability distributions of the input variables. The first d columns of the matrix are defined as matrix A and the remaining d columns are defined as matrix B. d further $N \times d$ matrices A_B^i, for $i = 1, 2, ..., d$, can be constructed so that the i-th column of A_B^i is equal to the i-th column of B, and the remaining columns are from A.

To estimate the main and total effect indices, two commonly used Monte Carlo estimators [Sob01][SAA+10] are introduced:
the estimator for the main effect index:

$$S_i = \frac{V_i}{Var(Y)} \approx \frac{\frac{1}{N}\sum_{j=1}^{N} f(B)_j (f(A_B^i)_j - f(A)_j)}{Var(Y)} \tag{2.62}$$

and the estimator for the total effect index:

$$S_{Ti} = \frac{E_{X_{\sim i}}(Var_{X_i}(Y \mid X_{\sim i}))}{Var(Y)} = \frac{\frac{1}{2N}\sum_{j=1}^{N}(f(A)_j - f(A_B^i)_j)^2}{Var(Y)} \tag{2.63}$$

2.4.3 Sampling Techniques

Several sampling techniques have been proposed for the uncertainty and sensitivity analysis. In this dissertation the simple random sampling and Sobol sequence techniques are implemented.

Simple random sampling Simple random sampling is a basic type of sampling and the principle is that each object has the same probability of being selected. According to the original probability density functions (PDFs) of the population (input variables), this technique can generate a random sample that satisfies the specified PDFs, which is suitable for cases where the statistical quantile with a confidence interval and also the tolerance limits need to be calculated.

Sobol sequence Sobol sequences are an example of quasi-random sequences with low discrepancy and were originally introduced by Ilya M. Sobol [Sob67]. The sample (Sobol sequences) generated by this technique has optimal space-filling properties. Due to their non-random nature, Sobol sequences are not suitable for calculating statistical quantiles with a confidence interval or the tolerance limits. However, due to its optimal space-filling properties it is suitable for the cases of training samples for Kriging based surrogate models. In addition, as a type of low-discrepancy sequence, the Sobol sequence is used to generate the sample for the variance-based sensitivity analysis as a quasi-Monte Carlo method.

2.4.4   Surrogate Model

In this dissertation the main objective of building surrogate model is to find a fast-running substitution of the original computer model due to the requirement of thousands of runs. The surrogate model is usually trained using data from actual simulations performed with the original computer model. About 100 runs are required to calculate the tolerance limits and these simulation results can be used to train the surrogate model.

However, to achieve a good space-filling property, coupled runs based on the sample generated by the Sobol sequence, which can better fill the space, are performed for the surrogate model.

As stated by O'Hagan [O'H06] Gaussian process regression (Kriging) is suitable for smooth response surfaces, which is the case of the computer model using the coupled code. The basic idea of Kriging, which can be treated as a form of Bayesian inference [Wil98], is to predict the value of a function at a given point by computing a weighted average of the known values of the function in the neighborhood of that point. Kriging starts with a prior distribution over functions. This prior takes the form of a Gaussian process: a sample of size N follows a normal distribution, where the covariance of any two sample points is the covariance function of the Gaussian process evaluated at the spatial location of two points. Thus, the value of a new point can be predicted by combining the Gaussian prior with a Gaussian likelihood function for each sample point. The resulting posterior distribution is also Gaussian, with a mean and covariance that can be easily computed from the observations, their variance, and the kernel matrix derived from the prior.

Spatial inference of a quantity $Z : \mathbb{R}^n \longrightarrow \mathbb{R}$ at an unobserved location x_0, is computed by a linear combination of the observations $z_i = Z(x_i)$ and the weights $\omega_i(x_0)$, $i = 1, ..., N$:

$$\hat{Z}(x_0) = [\omega_1 \ \omega_2 \ \cdots \ \omega_N] \cdot \begin{bmatrix} z_1 \\ z_2 \\ \vdots \\ z_N \end{bmatrix} = \sum_{i=1}^{N} \omega_i(x_0) \times Z(x_i) \tag{2.64}$$

where the weights ω_i are intended to summarize two important procedures in the spatial inference process:

- reflect the structural "proximity" of observed points to the estimation location x_0
- at the same time, they should have a desegregation effect, in order to avoid bias caused by eventual sample clusters

The two objectives in computing the weights ω_i are: unbias and minimum variance of the estimate. Several methods, such as ordinary Kriging and simple Kriging, have been developed to compute the weights under these objectives. When using ordinary Kriging, the calculation of the weights leads to the solution of the Kriging system:

$$
\begin{bmatrix} \hat{W} \\ \mu \end{bmatrix} = \begin{bmatrix} Var_{x_i} & 1 \\ 1^T & 0 \end{bmatrix}^{-1} \cdot \begin{bmatrix} Cov_{x_i x_0} \\ 1 \end{bmatrix} = \begin{bmatrix} \gamma(x_1, x_1) & \cdots & \gamma(x_1, x_N) & 1 \\ \vdots & \ddots & \vdots & \vdots \\ \gamma(x_N, x_1) & \cdots & \gamma(x_N, x_N) & 1 \\ 1 & \cdots & 1 & 0 \end{bmatrix}^{-1} \begin{bmatrix} \gamma(x_1, x^\star) \\ \vdots \\ \gamma(x_N, x^\star) \\ 1 \end{bmatrix}
$$

(2.65)

where the estimated weights vector $\hat{W} = \begin{bmatrix} \omega_1(x_0) \\ \omega_2(x_0) \\ \vdots \\ \omega_N(x_0) \end{bmatrix}$, the semivariogram $\gamma(x_i, x_j)$

$= \gamma(Z(x_i), Z(x_j))$, and μ is a Lagrange multiplier used in minimizing the Kriging error $\sigma_k^2(x)$ to satisfy the unbiasedness condition.

Based on the estimated weights $\hat{W}$, Kriging gives the best linear unbiased prediction of the intermediate values. Since it is beyond the scope of this document to investigate the mathematical algorithms of Kriging methods, further details are not discussed here. Thanks to the DACE package, developed by Lophaven et al. [LNS02], the implementation of Gaussian process regression can be performed in MATLAB and integrated into an a MATLAB-based uncertainty and sensitivity analysis platform.

2.4.5   Chain of Calculations

This section introduces the chains of calculations for forward and inverse uncertainty propagation.

Chain of Calculations of Forward Uncertainty Propagation

The chain of calculations of forward uncertainty propagation is shown in Figure 2.7. At the beginning the uncertain parameters and their Probability Distribution functions (PDFs) are identified and then a sample is generated accordingly. Based on the obtained sample, the input files for the uncertainty runs are generated. For each run, the steady-state simulations have to be performed for each code/field before the transient simulations, since the consistent initial conditions, especially at the coupled interfaces, have to be fulfilled for the computational convergence. Then, the codes/fields are coupled and the transient simulations are performed based on the achieved initial conditions in a parallel environment for a high computational efficiency. Once the transient simulations are finished, the data of the variables to be analyzed are extracted and post-processed, for the visualization with tolerance limits and further analysis.

Chain of Calculations of Inverse Uncertainty Propagation

The chain of calculations of inverse uncertainty propagation is shown in Figure 2.8. First, a high-fidelity simulation or experiment is performed to generate the reference data. Then the uncertain parameters are identified. For the parameters whose characteristics are well known, their PDFs are determined based on the available knowledge. For the parameters, whose characteristics are unknown, their PDFs are inferred with variables. The variables are obtained through an advanced constrained optimization that provides a minimum uncertainty range, ensuring that the tolerance limits with this range are able to cover the reference data. This obtained Probability Distribution function (PDF) has to be analyzed and a reasonable physical meaning should be applicable, if not, the form of the inferred PDF has to be modified accordingly and the constrained optimization is performed again until a PDF with reasonable physical meaning is delivered. Finally, all the uncertain parameters are summarized for further utilization.

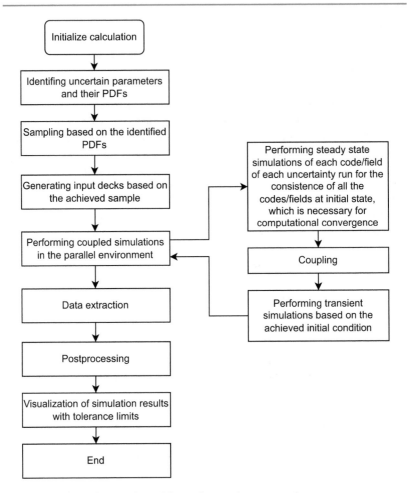

Figure 2.7 Chain of calculations of forward uncertainty propagation

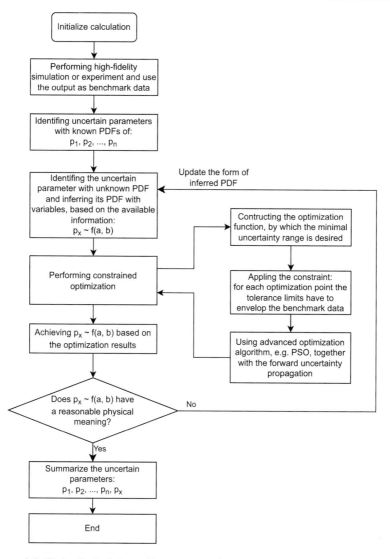

Figure 2.8 Chain of calculations of inverse uncertainty propagation

Model Verification

3

To ensure that the numerical models are capable of predicting the system behavior with sufficient accuracy, the models used in this dissertation have to be verified before further analysis.

The multiscale model of the coupled ATHLET-Ansys CFX code system has been validated against the experimental data and can be found in Chapter 4 for further details.

3.1 Verification of the CFD Model in COMSOL

Since the SMDFR design is a new concept and no experimental data is available as a reference, its computer model should be thoroughly verified. Since the model built in Ansys CFX is well validated by the TALL-3D facility [GPL+20], the CFD model in COMSOL can be verified against the model built in Ansys CFX. In addition, the solution was verified by the wall resolution in viscous units for the first layer, and a consistency of results obtained by models with different mesh scenarios was achieved for the mesh sensitivity study. Finally, it is shown that the CFD model built in COMSOL Multiphysics can be employed with a sufficient reliability for the thermal-hydraulic study.

3.1.1 Verification of the CFD Model in COMSOL against Ansys CFX

In order to obtain a reliable conclusion of the code-code verification, the turbulence models, geometries, physical setups (initial and boundary conditions), mesh

© The Author(s), under exclusive license to Springer Fachmedien Wiesbaden GmbH, part of Springer Nature 2023
C. Liu, *Multiscale and Multiphysics Modeling of Nuclear Facilities with Coupled Codes and its Uncertainty Quantification and Sensitivity Analysis*,
https://doi.org/10.1007/978-3-658-43422-9_3

scenarios, and boundary layers (Figure 3.1) were set to be the same for the model built in COMSOL Multiphysics and Ansys CFX. As shown in Figure 3.2, although there is an obvious discrepancy of temperature at the axial location of 0 to 0.4 m, for the remaining part, two models have the similar magnitude and tendency.

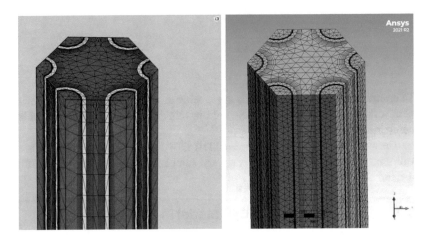

Figure 3.1 Mesh of the model in COMSOL (left) and Ansys CFX (right)

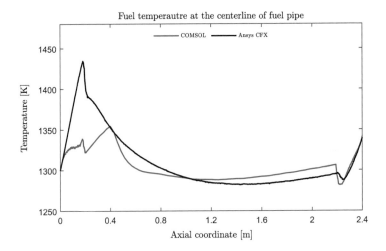

Figure 3.2 Fuel temperature at the centerline of fuel pipe

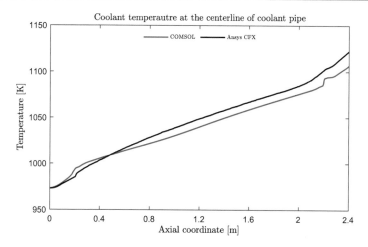

Figure 3.3 Coolant temperature at the centerline of coolant pipe

In addition, a consistency of the exit temperature (z = 2.4 m) is observed. Since this work aims at the investigating the system behavior, the accuracy of the fuel temperature calculated by the CFD model in COMSOL Multiphysics is considered sufficient.

The comparison of the coolant temperature is shown in Figure 3.3. The discrepancies between the results of the two codes are within ±15 K, which means that the coolant temperature calculated by the CFD model in COMSOL Multiphysics is consistent with that in Ansys CFX.

3.1.2   Verification of the CFD Model in COMSOL by Viscous Unit

The value of the wall resolution in viscous units indicates how far into the boundary layer the computational domain starts and should be less than 500 to ensure that the logarithmic layer meets the viscous sublayer. As shown in Figure 3.4, for more than 90% of the fluid domains the value of the wall resolution in viscous units lays between 11.6 and 100, which means that the first boundary layer corresponds approximately to the beginning of the logarithmic layer and the calculation in the first boundary layer has been performed correctly. No further refinement of the boundary layer mesh is necessary.

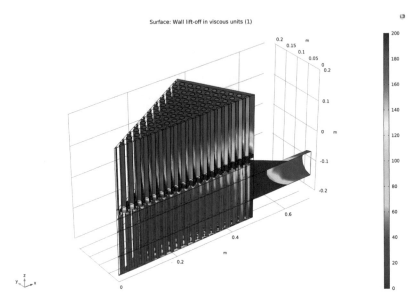

Figure 3.4 Wall lift-off in viscous units [1] at the wall surface

3.1.3   Verification of the CFD Model in COMSOL by Meshing Sensitivity

To investigate the mesh sensitivity, five mesh scenarios (M0 to M4) were selected, where M0 was the original mesh with six boundary layers, M1 and M2 had three and eight boundary layers respectively, and M3 and M4 have coarse and refined meshes separately. The mesh parameters are listed in Table 3.1. In order to quantify the simulation results for comparison and analysis, the outlet surface of the distribution zone was divided into 19 channels for molten salt, and 20 channels for liquid lead, along the radial direction from the center to the periphery, as shown in Figure 3.5. The mass flow rate averaged velocity and temperature of each channel were selected as the quantities to be compared and analyzed.

As shown in Figures 3.6 and Figure 3.7, all the mesh scenarios provided consistent results for the velocity and temperature profiles of both molten salt and liquid lead. However, in the central regions (near the centerline/axis of the core) some divergences were observed for M1 (black) and M3 (green). In order to ensure a high enough accuracy in the whole domain, the mesh scenario M0 (red), which gave con-

Table 3.1 Mesh Scenarios

Parameters	M0	M1	M2	M3	M4
Element size	Original	Original	Original	Coarse	Fine
Number of boundary layers	6	3	8	6	6
Number of elements	599,333	502,525	545,028	515,312	825,333
Average element quality	0.6259	0.6179	0.6222	0.596	0.6758
Minimum element quality	4.447×10^{-5}	3.75×10^{-6}	3.686×10^{-6}	3.467×10^{-6}	9.606×10^{-6}

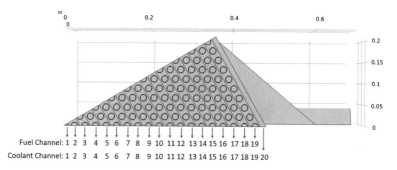

Figure 3.5 The channel indices of molten salt (fuel) and liquid lead (coolant) at the outlet plane

sistent results with M2 (magenta) and M4 (blue) but consumed less computational resources and was verified against the finer mesh scenario, was selected for all CFD models in COMSOL in this dissertation.

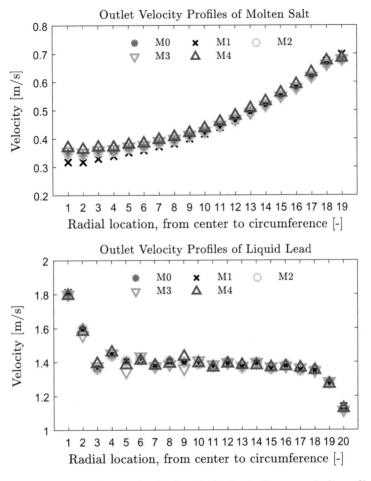

Figure 3.6 Mesh convergence study of outlet velocity distribution: top, velocity profiles of molten salt; bottom, velocity profiles of liquid lead

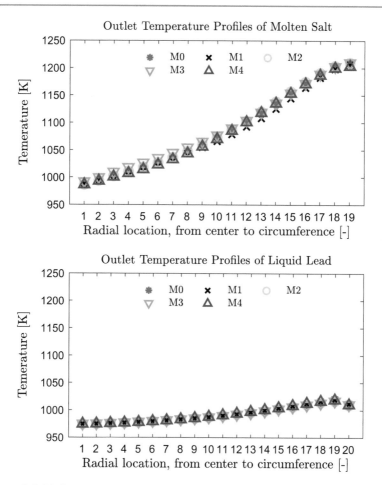

Figure 3.7 Mesh convergence study of outlet temperature distribution: top, temperature profiles of molten salt; bottom, temperature profiles of liquid lead

3.2 Verification of the Neutron Transport Model in COMSOL

For the single channel model, the neutron transport is solved directly by the user-defined diffusion equations. To verify its accuracy, three grouping scenarios were implemented: 1-group, 4-groups, and 8-groups. Based on these scenarios, the multiplication factor was calculated and compared with that obtained from the Serpent model as a reference, as shown in Table 3.2. The comparison of the neutron flux distribution is shown in Figure 3.8. It can be observed that all the group division scenarios gave consistent results for both the multiplication factor and the neutron flux distribution. Since the most focused phenomena in this dissertation are determined by the thermal-hydraulic behaviors, the 1-group model was selected for the steady-state and transient simulations of the single channel model.

Table 3.2 Comparison of the multiplication factor

Model types	Serpent	1-group	4-groups	8-groups
multiplication factor	1.03444	1.0394	1.0350	1.0362

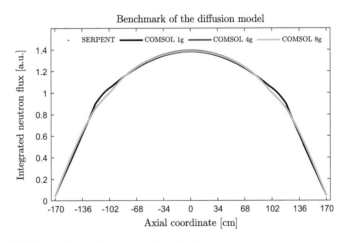

Figure 3.8 Comparision of the neutron flux distribution

Results and Analysis: SESAME TG03.S301.03　4

This chapter is divided into five parts: input uncertainties of the SESAME TG03.S301.03, the results in blind phase, the results in open phase, the surrogate model, and the sensitivity analysis.

4.1　Input Uncertainties of the SESAME TG03.S301.03

As listed in Table 1.1, the test TG03.S301.03 is the most challenging one due to flow instability and limit cycle oscillations propagating in the primary TALL-3D circuit. An interesting combination of two heating powers: 0.72 kW for MH, 10.3 kW for TS heater is selected by the experimentalists and kept constant during the whole transient, where the phenomena during forced to natural transient are investigated. As a representative. the following uncertainty and sensitivity analysis focuses on it. It is defined that the transient is initiated at 0 s due to pump tripping. Numerical simulations are performed using 1D-3D coupled and 1D STH codes, in semi-blind and open phases.

The uncertain parameters, which are selected from the input parameters for uncertainty quantification, are assumed to be uniformly distributed in certain ranges and are summarized in Table 4.1.

According to this table, the uncertain parameters come from:

- Roughness of pipes in three legs, common parts and Rotamass;
- Main Heater (MH) and Test Section (TS) heating power;
- Pressure form loss coefficients of MH leg, EPM pump and ROTAMASS;
- Heat transfer coefficients (HTCs) in three legs, of forced or natural convection to LBE, of the insulation from outer wall to air, of ball valves and EPM pump;

C. Liu, *Multiscale and Multiphysics Modeling of Nuclear Facilities with Coupled Codes and its Uncertainty Quantification and Sensitivity Analysis*, https://doi.org/10.1007/978-3-658-43422-9_4

Table 4.1 Uncertain Parameters of the SESAME TG03.S301.03

#	Parameters	Input of Code	Min.	Max.
1	Roughness of pipes in MH leg (value)	ATHLET	4.10E-5	5.04E-05
2	Roughness of pipes in common parts (value)	ATHLET	4.10E-05	5.04E-05
3	Roughness of pipes in HX leg (value)	ATHLET	4.10E-05	5.04E-05
4	Roughness of pipes in TS leg (value)	ATHLET	4.10E-05	5.04E-05
5	Roughness of pipes in Rotamass (value)	ATHLET	4.10E-05	5.04E-05
6	Power of the MH (factor)	ATHLET	0.977	1.023
7	Power of the TS heater (factor)	ATHLET, CFX	0.985	1.015
8	PFL coefficient in MH leg (factor)	ATHLET	0.813	1.231
9	PFL coefficient of EPM pump (factor)	ATHLET	0.65	1.35
10	HTC in common parts (factor)	ATHLET	0.9	1.1
11	HTC in HX leg (factor)	ATHLET	0.9	1.1
12	HTC in MH leg (factor)	ATHLET	0.9	1.1
13	HTC in TS leg (factor)	ATHLET	0.9	1.1
14	Initialization temperature at TC1.0000 (additive)	ATHLET	−2.2	2.2
15	LBE dynamic viscosity (multiplicative)	ATHLET, CFX	0.92	1.08
16	LBE thermal conductivity (multiplicative)	ATHLET, CFX	0.85	1.15
17	HTC forced or natural convection to LBE (value)	ATHLET	0.85	1.15
18	Room temperature bottom part (additive)	ATHLET, CFX	−2.2	2.2
19	HTC insulation outer wall to air (value)	ATHLET, CFX	2.1	8.2
20	HTC ball valves (value)	ATHLET	2.1	8.2
21	HTC EPM pump (value)	ATHLET	2.1	8.2
22	LBE heat capacity (factor)	ATHLET	0.95	1.05
23	PFL coefficients in ROTAMASS (factor)	ATHLET	0.9	1.1

- Initialization temperature at TC1.0000, room temperature (ambient temperature) at the bottom part;
- Thermophysical properties of the LBE;

To perform the uncertainty and sensitivity analysis, tolerance limits are calculated using the following procedure:

- Generate a sample ($N = 93$) of uncertain parameters (Table 4.1) using the simple random sampling technique;
- Generate input files for the coupled code using the generated sample;
- Execute the runs using the generated input files in a parallel environment on a Linux cluster;
- Extract and post-process the simulation results in order to make the output variables machine-readable;
- Calculate the tolerance limits.

Based on the obtained tolerance limits the uncertainty analysis is then performed and surrogate models can be trained based on the simulation results. Once the surrogate model is built and verified, the sensitivity analysis can be performed using the variance-based analysis technique.

4.2   SESAME TG03.S301.03 Blind Phase

In the blind phase the experimental data for only the first 1000 s were provided by KTH and the participants had to provide the simulation results for 15000 s obtained by the coupled code system. After collecting the results in the blind phase, the experimental data for the whole experiment were provided to the participants (open phase).

Since the main purpose of this test is to investigate the natural circulation after the pump trip, the mass flow rates with tolerance limits (0.95, 0.95) in three legs (the MH leg, the TS leg and the HX leg) during the whole transient are plotted and compared with the experimental data (measurements). For each leg, the physical phenomena were analyzed using the temperature distributions, including the inlet and outlet temperatures of these three key components (the MH, the TS and the HX).

According to the deviation between simulation and experimental data, ideas for further modification of the computer model were proposed in the open phase to improve the simulation results. Finally the importance of uncertainty analysis, in this case the calculation of tolerance limits, based on the observed physical phenomena, was given.

In Figures 4.1 to 4.3 the mass flow rates in three legs are plotted with tolerance limits. As shown in the legend, the black curve denotes the best-estimated results, the blue and green curves denote the upper and lower tolerance limits, respectively, and the red dots denote the experimental data (measurements).

Main Heater Leg
In the MH leg the measured mass flow rate during the transient has very low values of about 0.05 kg/s, which is only about 10% of the other two legs, and an oscillation is observed. Comparing the heating power and mass flow rates in the MH and TS legs, the mass flow rate driven by natural circulation in each leg is approximately proportional to its heating power. It can be considered as a competition of taking flow from HX leg by using the heating power between MH leg and TS leg.

The power of MH is quite low compared to the TS heater and HX (less than 10%). The MH leg can be treated as a vertical pipe, which is connected to a hot point and a cold point of a circuit composed by the TS leg and the HX leg (Figure 1.2, left). The fluid (LBE) in the MH leg is, to some extent, fixed by the pressure equilibrium, which is determined not only by the function of the temperature/density distribution in the loop, but also by the local pressure loss in the different components and by the buoyancy effects that take place. However, the heat sources in both the MH and

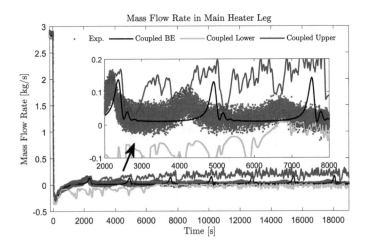

Figure 4.1 Mass Flow Rate in Main Heater Leg Blind Coupled

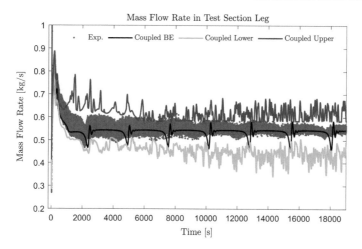

Figure 4.2 Mass Flow Rate in Test Section Leg Blind Coupled

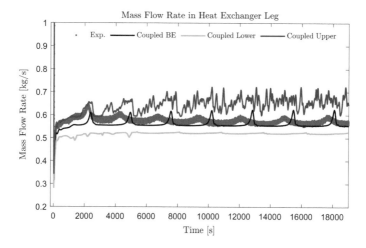

Figure 4.3 Mass Flow Rate in Heat Exchanger Leg Blind Coupled

the TS legs heat the fluid and then periodically unbalance the pressure equilibrium by perturbing the temperature distribution in the loop. The heating powers and volumes of the two components (the MH and the TS) have a combined influence on this behavior. When the pressure equilibrium is perturbed at several specific points ($t = 2200$ s, $t = 4200$ s, $t = 7000$ s, et al.), the fresh cold fluid coming from the bottom of the circuit flows into the MH leg (Figure 4.1) and then the pressure equilibrium is regenerated. The creation and destruction of the pressure equilibrium are periodically repeated, which results in Limit Cycle Oscillations (LCOs) of the mass flow rate in the MH leg. As a consequence of the temperature redistribution, there is an internal 3D interaction inside the MH leg, which has to be considered for the prediction of the LCOs. Unfortunately, since the entire MH leg is modeled by ATHLET, a system thermal-hydraulic analysis code, this local (non-axial) 3D pattern, which introduces more pressure losses and increases the heat transfer by natural convection, cannot be simulated. Therefore, the small-scale oscillations cannot be observed in the curve of the single best-estimate simulation (Coupled BE). However, the global behavior of this oscillation is predicted, and the total fluctuation is covered by the tolerance limits ("Coupled Lower" and "Coupled Upper"), which means that the use of the tolerance limits provides a better prediction compared to the standalone best-estimate simulation.

The outlet and inlet temperatures of MH are shown in Figure 4.4 and Figure 4.5, respectively. Similar to the mass flow rate, an oscillation is observed at the outlet. The magnitude of the temperature at the outlet is well predicted, while there is still a phase shift and the period of the oscillation is overestimated. However, the overall temperature trend is within the tolerance limits, which means that the uncertainty analysis was successful in predicting the temperature range during the entire transient.

To further investigate the temperature distribution in the MH leg, six figures for six time points are generated in Figure 4.6, where the red dots denote the experimental measurements and the black dots with error bars denote the simulation results with tolerance limits. Six time points from $t = 3000$ s to $t = 5500$ s are selected to illustrate the temperature distribution at different transient stages in the MH leg. As shown in Figure 1.1, location 1–7 denote three groups of thermocouples from bottom to top at MH leg:

- location 1–2: TC1.0000 and TC1.0346 at the bottom and below the MH;
- location 3–5: TC1.1740, TC1.2140 and TC1.2641 at the outlet of the MH and in the middle of the MH leg;
- location 6–7: TC1.4990 and TC1.5830 at the top.

At $t = 3000$ s, a large temperature difference between location 2 and location 3 is observed, which means that the temperature gradient inside the MH is quite large. Above the outlet of the MH, the temperature decreases along with height, except for location 6, where the pressure equilibrium is not finally established. After 500 s,

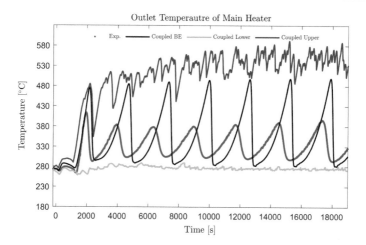

Figure 4.4 Outlet Temperature of Main Heater Blind Coupled

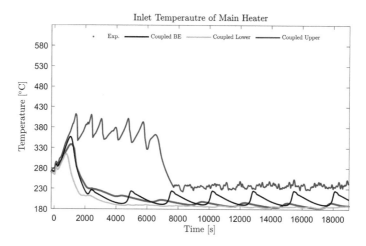

Figure 4.5 Inlet Temperature of Main Heater Blind Coupled

at $t = 3500$ s, due to the thermal conduction, the pressure equilibrium is finally established and the temperature distribution is temporarily kept stable. As shown in Figure 4.6, the temperature of the fluid is continuously decreasing above the MH, where a temperature gradient (distribution) is observed. For this particular moment the fluid can be considered as stagnant due to the density lock effect, which is consistent with the phenomenon observed in Figure 4.1.

At $t = 4000$ s, the temperatures at positions 4–5 are increased to the magnitude of the temperature at position 3 due to the heating of the MH. This changes the temperature distribution and the pressure equilibrium doesn't hold as before. This process eventually leads to the breaking of the fluid stagnation and a mass flow rate is observed for the moment, which is consistent with the phenomenon observed in Figure 4.1. Due to the injection of fresh cold fluid, the hot fluid inside the MH leg is partially partly flushed out. Therefore, the temperatures at positions 3–5 are reduced at $t = 4500$ s. When the MH is occupied by cold fluid, the fluid in the MH leg becomes stagnant again and the establishment of pressure equilibrium (density lock) is initiated again, which can be deduced from the temperature distributions at $t = 5000$ s and $t = 5500$ s.

Although the temperature evolution is generally predicted by the simulation, a significant overestimation of the magnitude is observed at $t = 4500$ s compared to the experimental data. Since in natural circulation the fluid flow is driven by the buoyancy force, a larger temperature gradient results in a larger buoyancy force. Considering the low power of MH, the resulting complex physical phenomena are quite difficult to predict correctly and are sensitive to the boundary conditions. As a 1D code, the ATHLET code is not good at capturing the evolution of the temperature distribution compared to the 3D code. However, the temperature distributions of the entire transient are covered by the tolerance limits, which means that by considering the uncertainty, the coupled model is able to predict the system behavior.

Test Section Leg

In the TS leg, the measured mass flow rate during the transient has a relatively high value of about 0.5 kg/s, which is almost equal to the mass flow rate in the HX leg. Unlike the HX leg, a fluctuation is observed, which can be explained by the 3D buoyant flow in the TS and the heat and mass transfer with the MH leg. Even the flow fluctuation is not predicted, by calculating the tolerance limits the high frequency oscillations are not resolved but the low amplitude and low frequency oscillations are resolved.

The outlet and inlet temperatures of TS are shown in Figure 4.7 and Figure 4.8, respectively. Similar to the mass flow rate, a temperature oscillation is observed at the outlet. Since the magnitude of the heating power of the TS heater is much

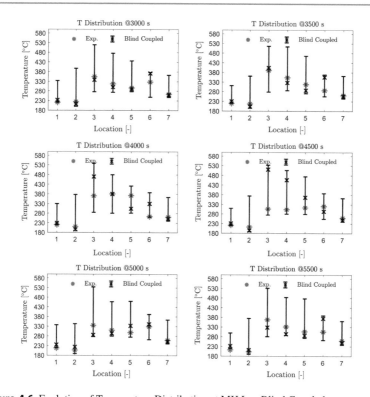

Figure 4.6 Evolution of Temperature Distribution at MH Leg Blind Coupled

higher than that of the MH, the natural circulation is relatively stable and thus the oscillation is not as significant. Since the peak temperature is overestimated, the heat loss of TS may be underestimated. In addition, the period of oscillation is slightly overestimated, which may be due to the overestimated friction in the TS leg. The whole temperature trend cannot be covered by tolerance limits, which means that the deviation is not due to the uncertainty. The physical effect causing it needs to be identified in the open phase for a better prediction.

To further investigate the temperature distribution inside TS, the values of CIP (circular inner plate TCs), BP (bottom plate TCs) and IPT (inner pipe TCs) at forced ($t = -100$ s) and natural ($t = 8000$ s) circulation stages are shown in Figure 4.9 and Figure 4.10, respectively.

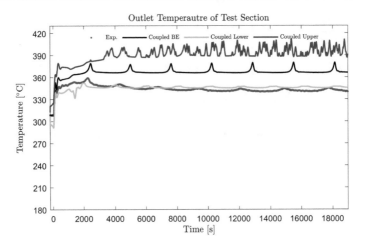

Figure 4.7 Outlet Temperature of Test Section Blind Coupled

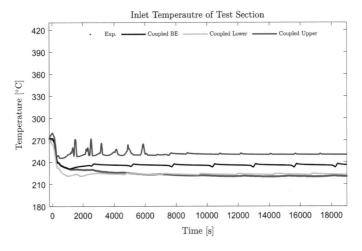

Figure 4.8 Inlet Temperature of Test Section Blind Coupled

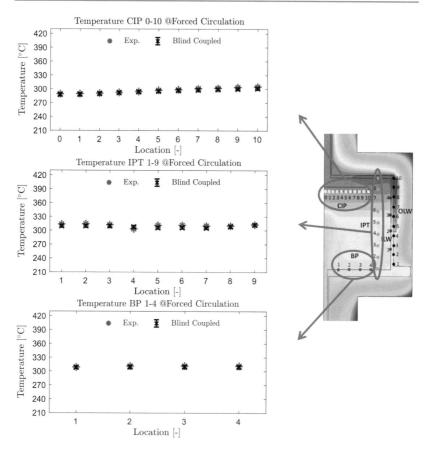

Figure 4.9 Temperature Distribution of TS @ Forced Circulation Blind Coupled

As shown in Figure 4.9, the temperature distribution during the forced circulation stage is well predicted. Due to the high mass flow rate, the strong thermal convection ensures that the fluid is well mixed during the heating process and no significant stratification occurs.

On the other hand, during the natural circulation stage, due to the low mass flow rate, the thermal convection is weak and the thermal conduction plays a non-negligible role, resulting in a significant stratification in the TS. Due to the complex local circulation, it is quite challenging to accurately capture the physical

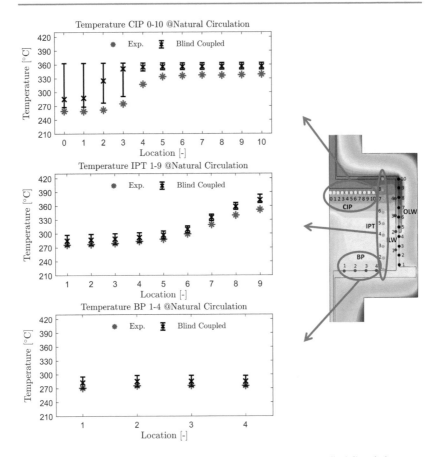

Figure 4.10 Temperature Distribution of TS @ Natural Circulation Blind Coupled

phenomena. Although the trend of the temperature distribution is generally predicted, the overestimation is so significant that the tolerance limits are not able to cover the experimental measurements. The inlet "jet" seems to reach a higher height in the experiment than in the calculation. This may be caused by the underestimation of the heat loss and the dynamic viscosity of the LBE and the overestimation of the thermal conductivity of the LBE, which can be verified by sensitivity analysis. Based on these facts, a modification of the heat loss coefficient and the friction factor in the open phase is considered, and the underlying uncertainty of the LBE properties has to taken into account. A more accurate quantification of these properties should be considered in future experiments.

Heat Exchanger Leg

In the HX leg, the measured mass flow rate at natural circulation is about 0.5 kg/s, which is almost equal to the mass flow rate in the TS leg. Unlike for the other legs, no fluctuation is observed because there is no significant 3D effect in this leg. Thus, the behavior is well predicted by the 1D code, and the measurements are generally within the tolerance limits.

As shown in Figure 4.11 and Figure 4.12, the temperature is also overestimated due to the interactions with the other legs. The tolerance limits do not cover the experimental measurements. In a way this is understandable, since the temperature overestimation in the MH and TS legs will lead to the temperature overestimation in the HX leg.

In Figure 4.12, a significant temperature fluctuation is observed, which means that the heat transfer between the primary circuit and secondary circuit in the HX is quite unstable in natural circulation. Since the secondary circuit is modeled in a simplified way and the oil is replaced by water, the heat transfer in the HX cannot be predicted accurately enough to capture this temperature fluctuation. However, by selecting appropriate uncertain parameters, this oscillation is covered by tolerance limits in the open phase.

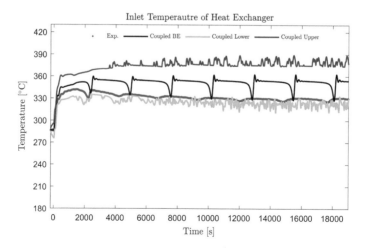

Figure 4.11 Inlet Temperature of Heat Exchanger Blind Coupled

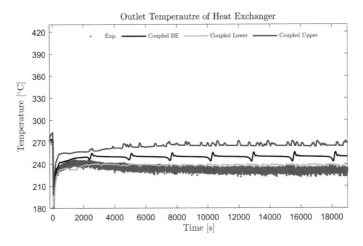

Figure 4.12 Outlet Temperature of Heat Exchanger Blind Coupled

Comparison with STH Code

To compare the prediction capabilities between the coupled code and the STH (Standalone Thermal-Hydraulics) code, a 1D simulation is performed by ATHLET using the same initial and boundary conditions, and the results obtained are compared with those obtained by the coupled code.

Figures 4.13 to 4.15 show the mass flow rates in three legs. As shown in the legend, the black curve denotes the best-estimated results, the blue and green curves denote the upper and lower limits respectively, the purple curve denotes the results from the STH code, and the red dots denote the experimental measurements as a reference. As shown in these figures, the results obtained by the 1D code are in agreement with those obtained by the coupled code, except for a slight phase shift. It is verified that the 1D code is capable of capturing the global behavior of the thermal-hydraulic system by properly modeling the TS.

Figure 4.16 to 4.20 show the outlet and inlet temperatures of MH, TS, HX with tolerance limits. As shown in the legend, the black curve denotes the best-estimated results, the blue and green curves denote the upper and lower limits respectively, the purple curve denotes the results from the STH code, the red dots denote the experimental measurements as a reference. Similar to the curves for the mass flow rates, the results obtained by the 1D code are in good agreement with those obtained by the coupled code, and only a small phase shift is observed. The 1D code's capacity of predicting the global system behavior is further verified.

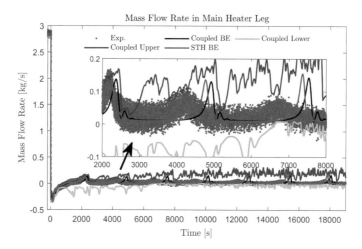

Figure 4.13 Mass Flow Rate in Main Heater Leg Blind Comparison

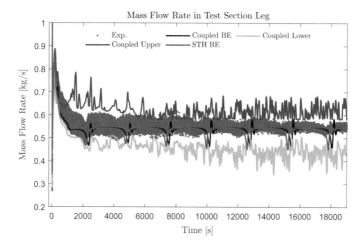

Figure 4.14 Mass Flow Rate in Test Section Leg Blind Comparison

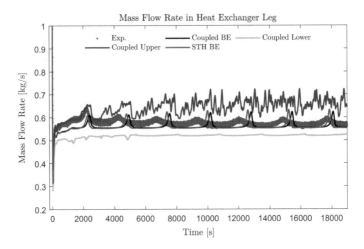

Figure 4.15 Mass Flow Rate in Heat Exchanger Leg Blind Comparison

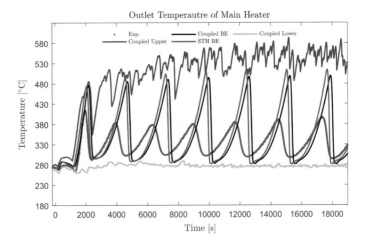

Figure 4.16 Outlet Temperature of Main Heater Blind Comparison

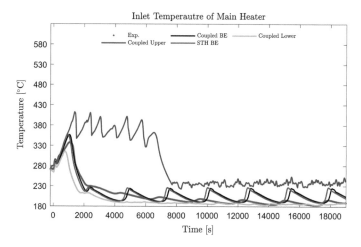

Figure 4.17 Inlet Temperature of Main Heater Blind Comparison

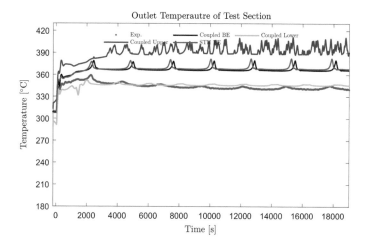

Figure 4.18 Outlet Temperature of Test Section Blind Comparison

4.3   SESAME TG03.S301.03 Open Phase

In the open phase, the experimental data for the whole transient are delivered and the computer model is updated based on the conclusions of Section 4.2. The pressure form losses in the electromagnetic pump, the heat transfer through the HX and the heat losses in the whole facility are modified at GRS. [PGL+19] The results with tolerance limits (0.95, 0.95) obtained with the modified model are compared with those obtained with the original model in blind phase and experimental data (Figure 4.21).

Since the main purpose of this test is to investigate the natural circulation after the pump trip, the mass flow rates with tolerance limits (0.95, 0.95) in three legs (MH leg, TS leg and HX leg) are plotted in Figure 4.22 to 4.24. As shown in the legend, the black curve denotes the best-estimated results from the updated model in the open phase, the blue and green curves denote the upper and lower limits respectively, the purple curve denotes the best-estimated results from the original model in the blind phase, and the red dots denote the experimental data as a reference. For each leg, the physical phenomena are described by temperature distributions, including the inlet and outlet temperatures of the three main components (MH, TS and HX).

In Figure 4.22, an increase of the lower tolerance limit can be observed at 4500 s, because the system reached a relatively steady state (started to stabilize) at 4500 s for a certain uncertainty run.

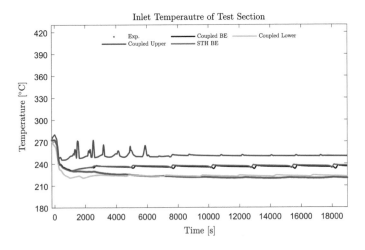

Figure 4.19 Inlet Temperature of Test Section Blind Comparison

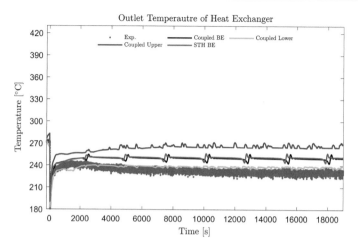

Figure 4.20 Outlet Temperature of Heat Exchanger Blind Comparison

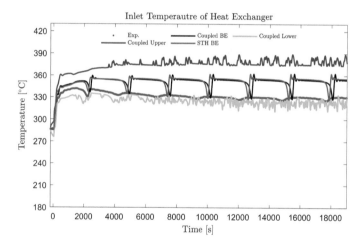

Figure 4.21 Inlet Temperature of Heat Exchanger Blind Comparison

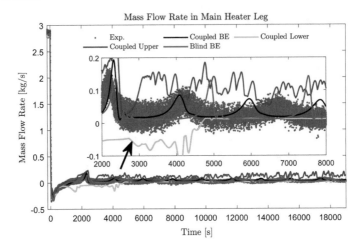

Figure 4.22 Mass Flow Rate in Main Heater Leg Open Coupled

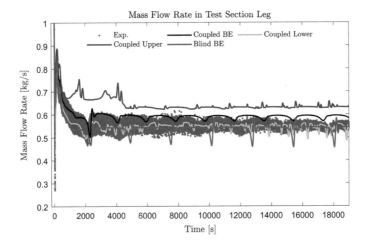

Figure 4.23 Mass Flow Rate in Test Section Leg Open Coupled

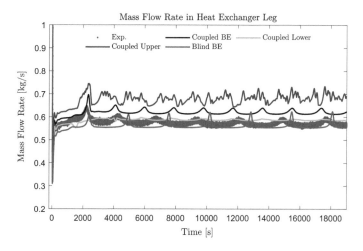

Figure 4.24 Mass Flow Rate in Heat Exchanger Leg Open Coupled

Main Heater Leg

As shown in Figure 4.22, the mass flow rates predicted by the modified model are lower and in better agreement with the experimental data. However, the oscillation period is underestimated.

The outlet and inlet temperatures of the MH are shown in Figure 4.25 and Figure 4.26, respectively. The temperature magnitude at the outlet is well predicted by the updated model, while the phase shift is still present and the oscillation period is underestimated. However, the whole temperature evolution is well covered by the tolerance limits as in the blind phase, and additionally the interval between the upper and lower limits is narrower. This means that the temperature evolution is more accurately predicted in the open phase.

To further investigate the temperature distributions of the MH leg, six figures are shown in Figure 4.27, where the red dots denote the experimental measurements and the black dots with error bars denote the simulation results with tolerance limits. Six time points from 3000 s to 5500 s are selected to illustrate the temperature distribution at different transient stages in the MH leg. As shown in Figure 1.1, location 1–7 denote three groups of temperature couples from bottom to top at MH leg:

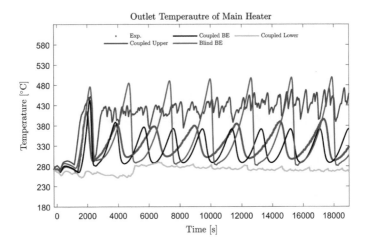

Figure 4.25 Outlet Temperature of Main Heater Open Coupled

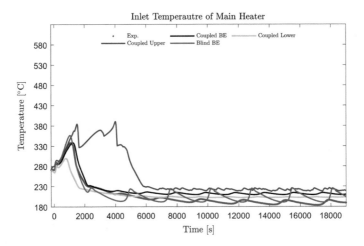

Figure 4.26 Inlet Temperature of Main Heater Open Coupled

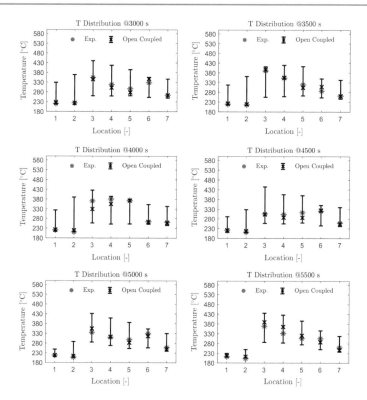

Figure 4.27 Evolution of Temperature Distribution at MH Leg Open Coupled

• location 1–2: TC1.0000 and TC1.0346 at the bottom and below the MH;
• location 3–5: TC1.1740, TC1.2140 and TC1.2641 at the outlet of the MH and in the middle of the MH leg;
• location 6–7: TC1.4990 and TC1.5830 at the top.

Compared to Figure 4.6, the evolution of the temperature distribution in the MH leg is more accurately predicted. The mass flow rate in this leg is so low that the friction losses are most likely not the main contributor to this behavior. At the same time, considering the improved prediction of mass flow rate, a close relationship between the evolution of the temperature distribution and the mass flow rate can be ensured, i.e. a better prediction of the temperature distribution leads to a better prediction of the mass flow rate, and vice versa.

Test Section Leg

The mass flow rate in the TS leg predicted by the modified model is slightly higher compared to the value predicted in the blind phase. The outlet and inlet temperatures of the TS are shown in Figure 4.28 and Figure 4.29 respectively. Due to the increased

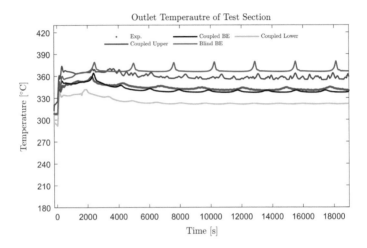

Figure 4.28 Outlet Temperature of Test Section Open Coupled

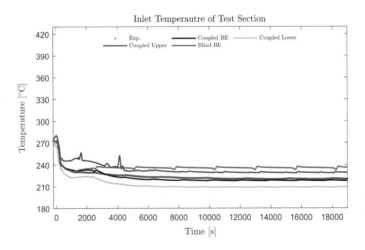

Figure 4.29 Inlet Temperature of Test Section Open Coupled

heat loss, the temperature prediction is improved and the tolerance limits are able to successfully cover the whole temperature trend.

To further investigate the temperature distributions within the TS, the values of CIP (circular inner plate TCs), BP (bottom plate TCs) and IPT (inner pipe TCs) at forced ($t = -100$ s) and natural ($t = 8000$ s) circulation stages are shown in Figure 4.30 and Figure 4.31, respectively.

As shown in Figure 4.30, the temperature distribution at the forced circulation stage is well predicted in both the blind and open phases. At the natural circulation

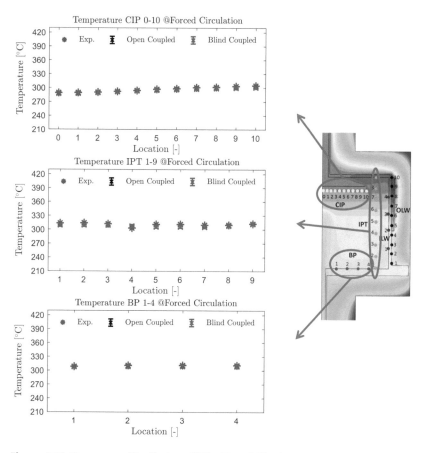

Figure 4.30 Temperature Distribution of TS at Forced Circulation Open Coupled

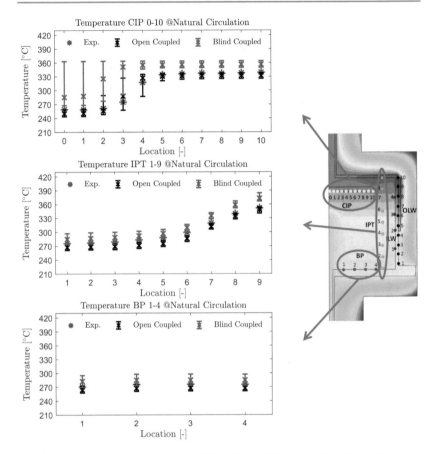

Figure 4.31 Temperature Distribution of TS at Natural Circulation Open Coupled

stage, the temperature distribution is generally well predicted in the open phase (Figure 4.31) due to the modification of the heat loss, while there are still some deviations. The tolerance limits cover the experimental measurements well, which means that the 3D local circulation inside the TS is generally well captured.

Heat Exchanger Leg

The mass flow rate in the HX leg predicted by the modified model is slightly higher compared to the value predicted in the blind phase due to the reduction of the form loss in the HX. As shown in Figure 4.32 and 4.33, the outlet and inlet temperatures of the HX are lowered due to the increased heat loss, thus improving the prediction in the open phase. The tolerance limits cover the experimental measurements to

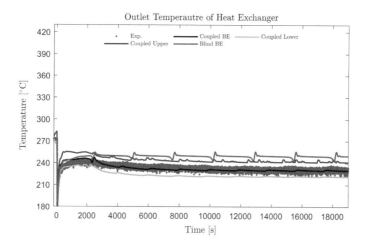

Figure 4.32 Outlet Temperature of Heat Exchanger Open Coupled

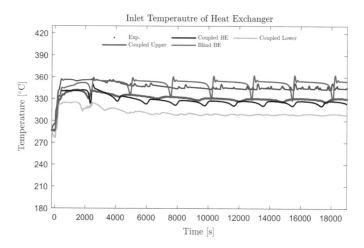

Figure 4.33 Inlet Temperature of Heat Exchanger Open Coupled

some extent, which means that the temperature evolution in the HX leg is generally
well predicted.

Comparison with STH Code
To investigate the difference in prediction capability between the coupled code and
the STH code (ATHLET standalone), a 1D simulation based on the STH code is
performed using the same initial and boundary conditions, and the results obtained
are compared with those obtained by the coupled code.

Figures 4.34 to 4.36 show the mass flow rates in three legs with tolerance limits.
As shown in the legend, the black curve denotes the best-estimated results, the blue
and green curves denote the upper and lower limits respectively, the purple curve
denotes the results by STH code, the red dots denote the experimental data as a
reference. As shown in these figures, the results obtained by the 1D code are in
agreement with those obtained by the coupled code, except for a slight phase shift.
It is verified that the 1D code is capable of capturing the global behavior of the
thermal-hydraulic system by properly modeling the TS.

In Figures 4.37 to 4.41 the outlet and inlet temperatures of MH, TS, HX are
plotted with tolerance limits. As shown in the legend, the black curve denotes the
best-estimated results, the blue and green curves denote the upper and lower limits
respectively, the purple curve denotes the results by STH code, the red dots denote
the experimental data as a reference. Similar to the curves of mass flow rate, the

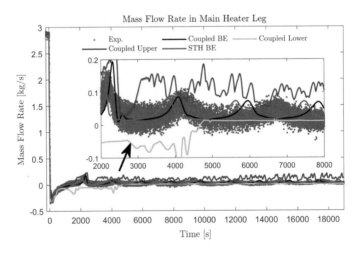

Figure 4.34 Mass Flow Rate in Main Heater Leg Open Comparison

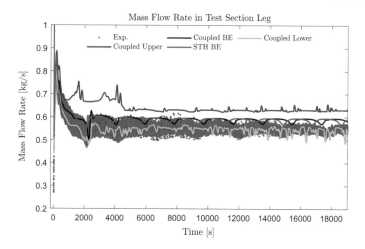

Figure 4.35 Mass Flow Rate in Test Section Leg Open Comparison

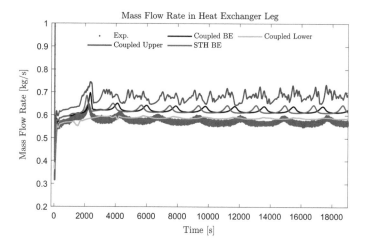

Figure 4.36 Mass Flow Rate in Heat Exchanger Leg Open Comparison

results of the 1D code and the coupled code are in agreement except for a slight phase shift (Figure 4.42).

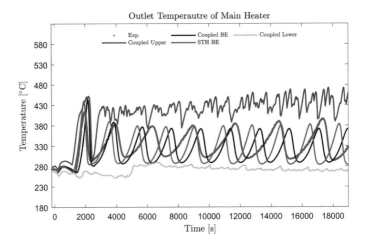

Figure 4.37 Outlet Temperature of Main Heater Open Comparison

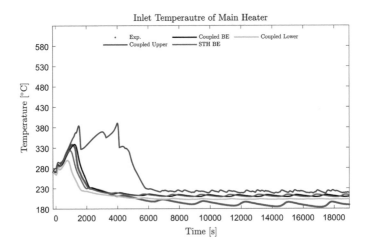

Figure 4.38 Inlet Temperature of Main Heater Open Comparison

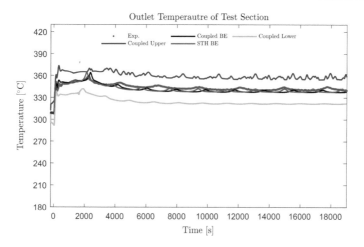

Figure 4.39 Outlet Temperature of Test Section Open Comparison

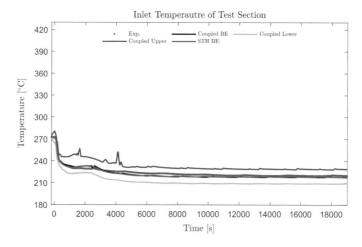

Figure 4.40 Inlet Temperature of Test Section Open Comparison

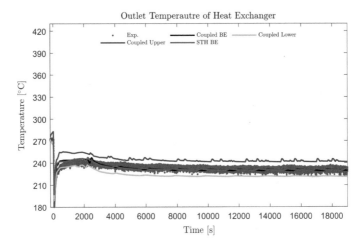

Figure 4.41 Outlet Temperature of Heat Exchanger Open Comparison

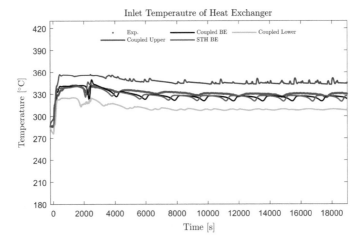

Figure 4.42 Inlet Temperature of Heat Exchanger Open Comparison

4.4   SESAME TG03.S301.03 Surrogate Model

For sensitivity analysis using the variance-based method, the number of runs is prohibitive. A surrogate model has to be built to make this number of runs affordable. 1000 runs based on Sobol sampling have been performed and the results obtained are suitable for training the surrogate model. The data from the first 800 runs are used to train the surrogate model and the remaining 200 runs are used to verify the accuracy of the surrogate model.

As an important variable, the mass flow rate in the TS leg at the natural circulation stage, which is defined by the averaged value from $t = 18000$ s to $t = 20000$ s, is selected as the output variable to be analyzed. Using the DACE package, 10 surrogate models based on different combinations of regression and correlation models are trained and verified. The results are summarized in Table 4.2.

The five correlation functions can be divided into two groups, one containing functions with parabolic behavior near the origin (Gaussian and Cubic spline), and the other containing functions with a linear behavior near the origin (Exponential, Linear and Spherical). As shown in Table 4.2, the linear group has a lower mean squared error compared to the parabolic group, so the system response of the TS leg mass flow rate can be considered linear near the given points, which is verified by the multiple determination coefficient calculation in Figure 4.43. In addition, since all the surrogate models provide a sufficiently low mean squared error, it is reasonable to select the Surrogate Model 6, which uses the first-order polynomial

Table 4.2 Surrogate Models

Surrogate Model	Regression Model	Correlation Model	Mean Squared Error
1	Zero-order polynomial	Exponential	1.5686E-06
2	Zero-order polynomial	Gaussian	2.1353E-06
3	Zero-order polynomial	Linear	1.5699E-06
4	Zero-order polynomial	Spherical	1.5524E-06
5	Zero-order polynomial	Cubic spline	2.6330E-06
6	First-order polynomial	Exponential	1.4962E-06
7	First-order polynomial	Gaussian	1.6059E-06
8	First-order polynomial	Linear	1.5612E-06
9	First-order polynomial	Spherical	1.5236E-06
10	First-order polynomial	Cubic spline	1.6214E-06

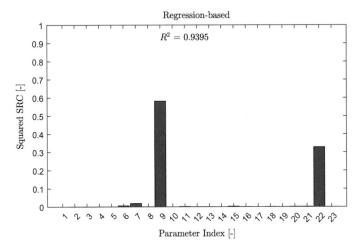

Figure 4.43 Parameter sensitivity on the averaged mass flow rate in the TS leg during natural circulation using regression-based method

regression model and the exponential correlation model, as the model used for the following sensitivity analysis: the parameter sensitivity to the mass flow rate as well as to other system responses.

4.5   SESAME TG03.S301.03 Sensitivity Analysis

In this section the mass flow rate in the TS leg at the natural circulation stage, which is defined by the averaged value from $t = 18000$ s to $t = 20000$ s, is selected as the variable to be analyzed. By performing the sensitivity analysis based on the regression-based technique and the variance-based technique separately, the importance level of all the 23 uncertain parameters (Table 4.1) is quantitatively evaluated.

Based on the regression-based technique, the input and output variables are standardized using Formula (2.47), and then the Person's Ordinary Correlation Coefficients (CCs) and the Spearman's Rank Correlation Coefficients (RCCs, based on the ranks) are calculated for each input variable (uncertainty parameter) using Formula (2.46). To verify the linearity of the system, the coefficient of multiple determination is calculated using Formula (2.48). Since the coefficient based on RCCs is 0.9395, which is closed to 1.0, a linear correlation is considered as the main part when evaluating the influence of input parameters on the mass flow rate at the natural circulation stage. However, since it doesn't explain 6% of the output variability, the calculation of Sobol indices (variance-based technique) is also performed.

As shown in Figure 4.44, Parameter 9, which is the pressure form loss coefficient of the EPM-pump (main pump) according to Table 4.1, has the largest influence (69%) on the output variable. This means that the calibration of the EPM-pump (characteristic curve, contributes to the evolution of the friction loss during the transient) should be the most important consideration when modifying the computer model. In addition, Parameter 22 (LBE heat capacity, contributes to the temperature distribution) has the second most important influence (29%) on the system output variable and should also be considered for calibration to further improve the prediction capability. The remaining parameters have almost no influence on the averaged mass flow rate in the TS leg during natural circulation and can be considered as having no contribution.

Based on the variance-based technique, samples of $N = 10000$ and $N = 100000$ are generated using the Sobol sequence technique. These samples are then entered into the surrogate model. Finally the first-order index $S1$ and the total-effect index ST are obtained using Formulas (2.62) and (2.63) for each input variable separately.

As shown in Figure 4.44, the results of all the samples are consistent, which means that a sample of $N = 10000$ is sufficient to generate the precise first-order index $S1$ and total-effect index ST. When comparing the first-order index $S1$ and the total-effect index ST, a slight deviation is observed. This is due to the fact that the system is not absolutely linear and that the 2-order and higher-order indices have to be included for the global sensitivity analysis. Furthermore, a general consistence is confirmed in comparison with the results obtained by the regression-based method.

Finally, two parameters are identified as influential by the global sensitivity analysis: the pressure form loss coefficient of the EPM pump (most important) and the heat capacity of the LBE.

Besides the mass flow rate, the influences of uncertain parameters on the amplitude, peak value, and frequency of the temperature oscillation in the MH leg are shown in Figure 4.45, Figure 4.46 and Figure 4.47.

For the amplitude and peak value, the LBE heat capacity is identified as the most influential parameter, since it has a deterministic contribution to the evolution of the LBE temperature distribution in the natural circulation stage. It is obvious that its importance level is underestimated by the regression-based method and the first-order index, since a large part of the nonlinear contribution and the co-contribution with other parameters are not included. In this case, the total-effect index is recommended as an indicator that can reflect the total contribution of the parameters. In addition, the PFL coefficient of the EPM pump (contributes to the

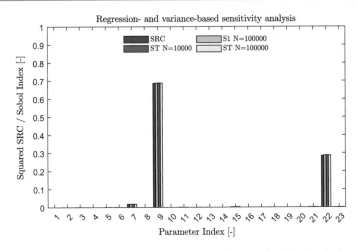

Figure 4.44 Parameter sensitivity to the averaged mass flow rate in the TS leg during natural circulation using regression- and variance-based method

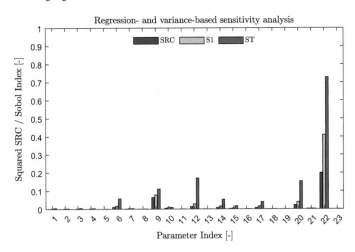

Figure 4.45 Parameter sensitivity to the oscillation amplitude of the MH outlet temperature using regression- and variance-based method

mass flow rate), the HTC in the MH leg (contributes to the heating power), the HTC ball valve (contributes to the heating power) have certain influences on the amplitude and peak value of the fuel temperature.

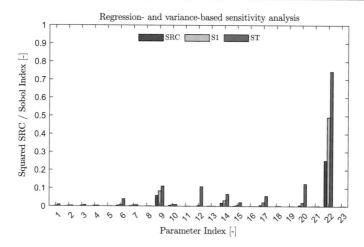

Figure 4.46 Parameter sensitivity to the peak of the MH outlet temperature using regression- and variance-based method

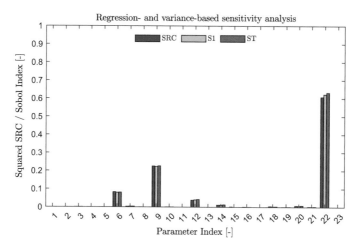

Figure 4.47 Parameter sensitivity to the frequency of oscillations of the MH outlet temperature using regression- and variance-based method

For the frequency, the LBE heat capacity is identified as the most influential parameter for the same reason as mentioned above. However, the importance level obtained by the regression-based method, by the first-order index and by the total-effect index is almost the same, which means that a linear and direct contribution is the major part. Besides, the PFL coefficient of the EPM pump (contributes to the mass flow rate) and the power of the MH (contributes to the heating power) have certain influences on the fuel temperature fluctuation frequency.

All the identified parameters should be considered for further calibration to improve the predictive capability. Since the risk of pipe failure is increased by the thermal shock caused by the temperature fluctuation during natural circulation, the uncertainties of these parameters should be reduced for a safety analysis with higher reliability.

Results and Analysis: SMDFR Single Channel **5**

This chapter is divided into five parts: input uncertainties of the SMDFR single channel model, the results of the SMDFR single channel model at steady state, its corresponding uncertainty and sensitivity analysis, the responses of the SMDFR single channel model at various transients, and its corresponding uncertainty and sensitivity analysis.

To quantify the system uncertainties, the input uncertainties of the system must be identified, and their ranges and probability density functions must be determined. Several methods are available for the quantification of the uncertainty, but the non-destructive technique, tolerance limits [Gla08], is selected in this dissertation considering its reduced consumption of computational resources.

5.1 Input Uncertainties of the SMDFR Single Channel Model

Regardless of the accuracy that the numerical model can achieve, since approximations and assumptions are essential in the modeling and calculation process, a certain degree of uncertainty is always expected and has to be well quantified to provide reliable results with adequate assurance. However, since not all of the uncertain parameters can be considered due to the limited computational resources and the limited knowledge, only the most representative uncertain parameters are considered based on the experience of a previous study [LPMJ20]. Their influences on the numerical model have to be quantified and considered during the design and optimization of the control system.

© The Author(s), under exclusive license to Springer Fachmedien Wiesbaden GmbH, part of Springer Nature 2023
C. Liu, *Multiscale and Multiphysics Modeling of Nuclear Facilities with Coupled Codes and its Uncertainty Quantification and Sensitivity Analysis*,
https://doi.org/10.1007/978-3-658-43422-9_5

The selected uncertain parameters, which are considered as the input of the uncertainty quantification procedure, were assumed to be uniformly distributed in their ranges and are given in Table 5.1. The ranges are determined as follows:

- Thermal power, mass flow rates, and inlet temperatures are specified with reference to the reported experimental uncertainty reported in the literature [GPL+20].
- The heat capacity of molten salt (fuel) is specified with reference to the value of liquid lead. Since no data are available for the molten salt, and there is a heat transfer process between the two fluids, a consistent uncertainty range is applied in order to achieve a conservative assumption for the heat transfer process. In the future, its value will be adjusted when experimental data of the molten salt are generated/available.
- The thermal conductivity of the molten salt (fuel) is assumed with an uncertainty range of ±5% from the default value. Since no data are available, this assumption is considered to be sufficiently conservative. As more data becomes available, the ranges will be updated accordingly.
- The density and dynamic viscosity of molten salt (salt) are specified with reference to the accuracy of measurement reported in the literature [DKRC75].
- The thermal properties of liquid lead (coolant): heat capacity, thermal conductivity, density, and dynamic viscosity, are specified according to the uncertainty bounds found in [FSA+15].

Table 5.1 Uncertain parameters of the SMDFR single channel model

#	Parameters	PDF	Min.	Max.
1	Fuel inlet temperature (additive)	Uniform	−2.2	2.2
2	Coolant inlet temperature (additive)	Uniform	−2.2	2.2
3	Mass flow rate of fuel (factor)	Uniform	0.95	1.05
4	Mass flow rate of coolant (factor)	Uniform	0.95	1.05
5	Heat capacity of fuel (factor)	Uniform	0.93	1.07
6	Thermal conductivity of fuel (factor)	Uniform	0.95	1.05
7	Density of fuel (factor)	Uniform	0.99	1.01
8	Dynamic viscosity of fuel (factor)	Uniform	0.963	1.037
9	Heat capacity of coolant (factor)	Uniform	0.93	1.07
10	Thermal conductivity of coolant (factor)	Uniform	0.85	1.15
11	Density of coolant (factor)	Uniform	0.99	1.01
12	Dynamic viscosity of coolant (factor)	Uniform	0.95	1.05
13	Thermal power (factor)	Uniform	0.98	1.02

To perform the uncertainty quantification, tolerance limits are calculated using the following procedure:

- Generate a sample ($N = 93$) of uncertain parameters (Table 5.1) using the simple random sampling technique;
- Generate input files for the coupled code using the generated sample;
- Perform the runs based on the generated input files in a parallel environment on Linux-Cluster;
- Extract and post-process the simulation results in order to make the output variables machine-readable;
- Calculate the tolerance limits.

Based on the obtained tolerance limits, the uncertainty analysis is then performed and surrogate models can be trained based on the simulation results. Once the surrogate model is built and verified, the sensitivity analysis can be performed using the variance-based analysis technique.

5.2 SMDFR Single Channel Steady State

In this section, although the SMDFR single channel model has been built and solved in a 3D geometry, as shown in Figure 5.1 and Figure 5.2, the results are processed and presented in a 1D (axial) manner with the same nodalization in Figure 2.2 to highlight the main global features.

The system behaviors are investigated in four aspects: Heat transfer characteristics, Hydraulic characteristics, Neutronics characteristics, Uncertainty analysis

5.2.1 Heat Transfer Characteristics

In Figure 5.3, it is observed that the node-averaged temperature of liquid lead as coolant increases continuously from the inlet to the outlet. For all the 12 nodes, the coolant continues to absorb heat from the fuel, resulting in a temperature increase from 973 K to 1100 K. The temperature of molten salt as fuel decreases for the first 4 nodes and then increases to 1309 K at the last node (outlet). However, the peak temperature has to be checked to identify the hot spot, which is shown in Figure 5.4. A similar trend is observed for the coolant, but with a higher magnitude up to 1100 K. The trend of the fuel peak temperature has a different profile, and the highest peak temperature occurs at node 2 and node 3, which is around 1350 K.

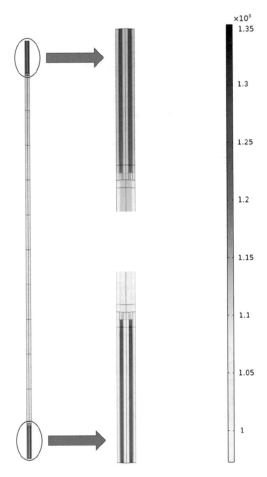

Figure 5.1 Temperature distribution in 3D color map (color legend unit in [K])

The temperature profile has to be analyzed with the fission power distribution and the heat transfer rate of the coolant, which are shown in Figure 5.5 and Figure 5.6. It is obvious that the fission power in the first node and the last node, as the distribution and collection zones, is much higher than in the other nodes because the fuel content occupies most of the volume in these two nodes due to the special geometry. The fission power distribution in node 2 to node 11 is similar to that of the conventional reactor type. Since both the temperature difference between the fuel and coolant

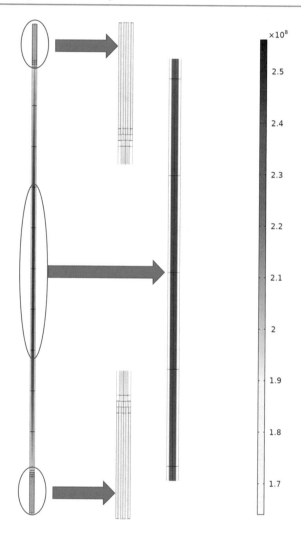

Figure 5.2 Power density distribution in 3D color map (color legend unit in $[W/m^3]$)

and the fission power are the largest in node 1, the heat transferred to the coolant has the highest value of about 18 kW, while the heat transfer rate in other nodes varies from 6 kW to 9 kW. For node 12, although the fission power is the largest, the heat absorbed by the coolant is not as high as that in node 1 due to the smaller temperature

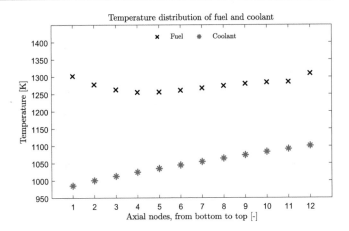

Figure 5.3 Temperature distribution of fuel and coolant at steady state

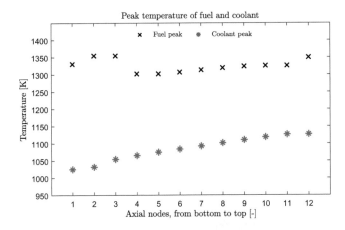

Figure 5.4 Peak temperature of fuel and coolant at steady state

difference. The coolant temperature increases from node 1 to node 12 because it continuously absorbs heat from the fuel continuously. For the fuel temperature, an energy balance between the generated fission power and the heat absorbed by the coolant has to be considered, and the temperature profile is relatively flat, which provides a lower risk of thermal shock to the SMDFR.

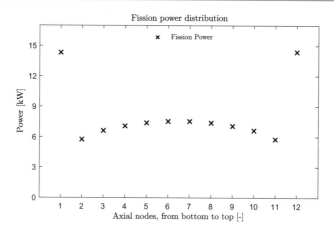

Figure 5.5 Fission power distribution

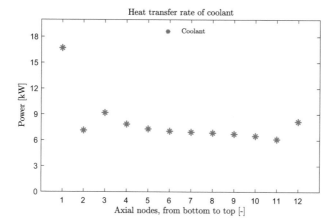

Figure 5.6 Heat transfer rate of coolant

The total heat absorbed by the coolant is 96.4 kW, which is at the same level as the total fission power of 97.1 kW. It has been demonstrated that with this design, the heat generated in the core can be successfully removed by the coolant at steady state.

5.2.2   Hydraulic Characteristics

To quantify the energy loss during the flow, Bernoulli's equation applied to a micro-element of fluid is introduced:

$$\frac{P_1}{\rho_1 g} + z_1 + \frac{u_1{}^2}{2g} = \frac{P_2}{\rho_2 g} + z_2 + \frac{u_2{}^2}{2g} + h_w{}' \tag{5.1}$$

multiplying both sides by a micro-flow element dm ($dm = \rho_1 g u_1 dA_1 = \rho_2 g u_2 dA_2$), and then integrating both sides by the corresponding cross-sectional areas A_1 and A_2:

$$\int_{A_1} (P_1 + \rho_1 g z_1) u_1 dA_1 + \int_{A_1} \frac{1}{2} \rho_1 u_1{}^3 dA_1$$
$$= \int_{A_2} (P_2 + \rho_2 g z_2) u_2 dA_2 + \int_{A_2} \frac{1}{2} \rho_2 u_2{}^3 dA_2 + h_w \tag{5.2}$$

where h_w is the energy loss in W, which determines the pumping power requirement. By selecting the inlet and outlet cross-sectional areas of the fluids as A_1 and A_2, the energy loss in the single channel was then obtained, which is equal to h_w. The h_w of fuel and coolant is shown in Figure 5.7, the pressure loss of fuel is minimal compared to that of coolant and mainly occurs at the first and last node due to the

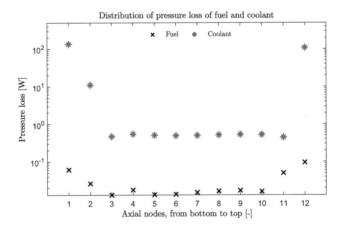

Figure 5.7 Distribution of pressure loss of fuel and coolant at steady state

sharp geometry change. Since the heat removal function is mainly performed by the liquid lead, the flow of molten salt is only required for chemical processing, and its mass flow rate does not need to be large. Due to its high mass flow rate, the h_w of liquid lead (for a single channel) is quite significant: 331 W, and it can be estimated that the total h_w of liquid lead in the core is about 340 kW. However, it should be noted that this single channel model is not capable of accurately representing the entire core behavior, and this estimate serves only as a qualitative overview. The coolant pump has to overcome at least this energy loss and should have provided more power considering the energy losses in other parts of the secondary circuit, especially in the heat exchanger.

5.2.3 Neutronics Characteristics

The neutron flux distribution is shown in Figure 5.8, which is similar to that of a conventional reactor. Despite the existence of the drift effect of the DNPs, the flux profile is not affected due to the low proportion of the delayed neutrons. However, this does not mean that the DNPs have no effect on the transient behavior. Since the delayed neutrons have a much longer generation time compared to the prompt neutrons, the DNPs play an important role in the system response under transient scenarios despite their small contribution to the neutron generation.

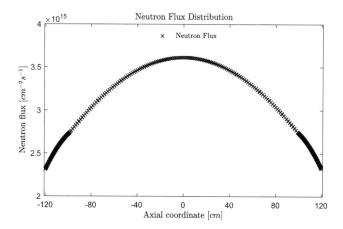

Figure 5.8 Neutron flux distribution

The Delayed Neutron Precursor (DNP) concentrations are shown in Figure 5.9 and Figure 5.10, in which there are six DNP groups: C_1 to C_6. Based on their concentration profiles, they can be divided into 2 categories: C_1 to C_3, and C_4 to C_6.

For C_1 to C_3, a steep rise is observed in both the distribution and colleaction zones (node 1 and node 12), where the generation rate is higher than in other nodes. Since the fuel content occupies most of the volume in these two nodes due to the special geometry, the increased total macroscopic fission cross section overcomes

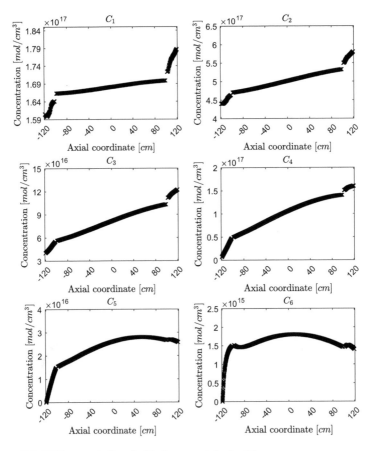

Figure 5.9 DNP concentrations inside the core at steady state

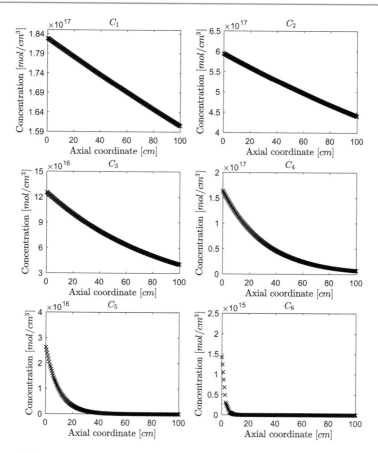

Figure 5.10 DNP concentrations outside the core at steady state

the lower neutron flux and results in a much higher fission rate, which means a higher generation rate of heat and DNPs. Due to their relatively low decay constants, the fission process has a dominant influence inside the core, and a linear shape can be used to describe the concentration distribution of DNPs. Similarly, the distribution of DNPs outside the core can be described by an exponential shape, where the decay process dominates. Before all the DNPs disappear by decay, they are sent back to the core for the next round of circulation.

For C_4 to C_6, the tendency of the concentration distribution is closer to the neutron flux distribution with a cosine-like shape. Due to their relatively large decay constants the decay process plays an important role in the concentration distribution. For the most obvious case C_6, the DNPs disappear in a short time after their production, which means that the concentration is proportional to the fission rate and its distribution is co-determined by the neutron flux distribution and the decay constant. Similarly, the distribution of DNPs outside the core can be described by an exponential shape, where the decay process is the only process determining the concentration. The DNPs disappear due to decay before they are sent back to the core, resulting in the concentration inside the core starting from zero.

5.3   Uncertainty and Sensitivity Analysis of SMDFR Single Channel Steady State Characteristics

5.3.1   Uncertainty Analysis of SMDFR Single Channel Steady State Characteristics

In order to quantify the uncertainties of the simulation results, an error bar was used to illustrate the two-sided tolerance limits: the length below the data point was determined by the lower limit and the length above the data point represented the upper limit.

As shown in Figure 5.11 and Figure 5.12, the uncertainty band of the average and peak temperature of the fuel is within ± 18 K, while the uncertainty of the coolant temperature increases from node 1 to node 12. Starting from the inlet, the uncertainty of the coolant temperature accumulates and reaches its maximum value at node 12, which is about $+14$ K$/-10$ K. This uncertainty range has to considered when performing a safety analysis, and it can be interpreted as follows: the average coolant temperature has a 95% probability of being between $T_{BE} + 14K$ and $T_{BE} - 10K$ with a confidence level of 95%, and a safety analysis should be able to demonstrate that the system is safe for any coolant temperature in this range.

Referring to Figure 5.13, a higher uncertainty band is observed at node 1 and node 12. And the coolant absorbs the largest amount of heat at node 1, as shown in Figure 5.14. These uncertainty results mean that node 1 and node 12 have the largest uncertainty and further research should focus on these two nodes. If available, more data on them have to be supplemented to narrow their uncertainty bands, which will effectively reduce the system uncertainties.

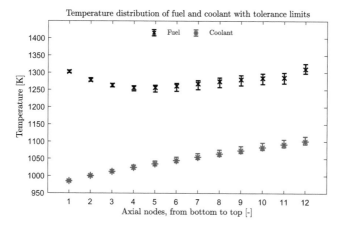

Figure 5.11 Temperature distribution of fuel and coolant with tolerance limits at steady state

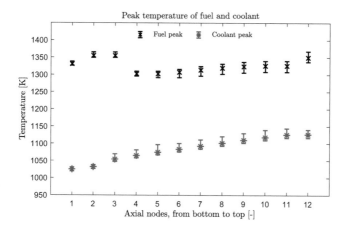

Figure 5.12 Peak temperature of fuel and coolant with tolerance limits at steady state

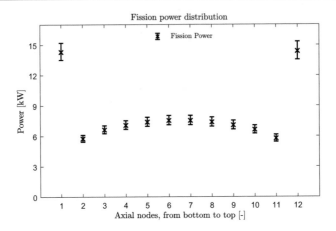

Figure 5.13 Fission Power Distribution with Tolerance Limits

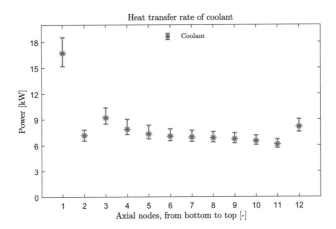

Figure 5.14 Heat Transfer Rate of Coolant with Tolerance Limits

5.3.2   Sensitivity Analysis of SMDFR Single Channel Steady State Characteristics

In order to quantify the influence of disturbances on the transient responses, the total fission power is selected as the output variable for the sensitivity analysis. By performing the sensitivity analysis based on the regression-based technique and the variance-based technique separately, the importance level of 13 uncertain parameters (Table 5.1) on the total fission power at steady state is quantitatively evaluated.

As shown in Figure 5.15, the coefficient of multiple determination is 0.7831 (higher than 0.6), which means that a linear correlation is considered to be the dominant part when evaluating the influence of uncertain parameters on the power. A general agreement between the regression-based method and the variance-based method is observed in Figure 5.16. Three parameters, Parameter 5 (heat capacity of fuel), Parameter 9 (heat capacity of coolant), and Parameter 10 (thermal conductivity of coolant), are identified as the most influential parameters with a greater contribution than 10%. This is because the heat capacities of two fluids and the thermal conductivity of the coolant have an influence on the temperature distributions and thus exert a reactivity feedback, resulting in a deviation of the fission power. In addition, Parameter 2 (coolant inlet temperature), Parameter 3 (mass flow rate of fuel), Parameter 4 (mass flow rate of coolant), Parameter 6 (thermal conductivity of fuel), and Parameter 12 (dynamic viscosity of coolant) have certain influences (less than 10%) on the fission power. All the identified parameters, especially the thermophysical properties of the fluids, should be further calibrated in order to reduce the uncertainties of the results and improve the prediction capability.

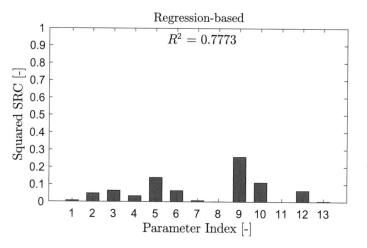

Figure 5.15 Parameter sensitivity to total fission power using regression-based method

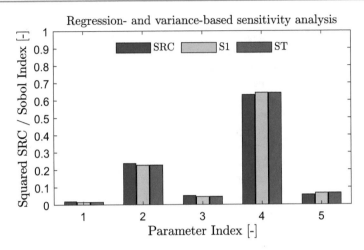

Figure 5.16 Parameter sensitivity to total fission power using regression- and variance-based method

---

5.4   SMDFR Single Channel Transient Responses

In this section several typical transients are simulated in order to investigate the system response under the following scenarios: reactivity insertion accident, variations of the fuel and coolant temperatures, variations of the fuel and coolant mass flow rates.

5.4.1   SMDFR Single Channel Transient: Reactivity Insertion Accident

In this reactivity insertion accident, a reactivity of 100 pcm was inserted uniformly throughout the channel. The temperature and power evolution of the 12 axial nodes are shown in Figure 5.17 to Figure 5.19. When the reactivity is introduced, the fuel temperature in all the nodes is increased in a short time (about 1 s) due to the increased power. When the fuel temperature is increased, due to the negative feedback of the temperature, a negative reactivity is introduced, and then the power starts to decrease, which results in a decrease of the fuel temperature. However, for node 7 to node 12, no temperature decrease occurs because the fuel temperature in these nodes is determined not only by the fission power, but also by the fuel

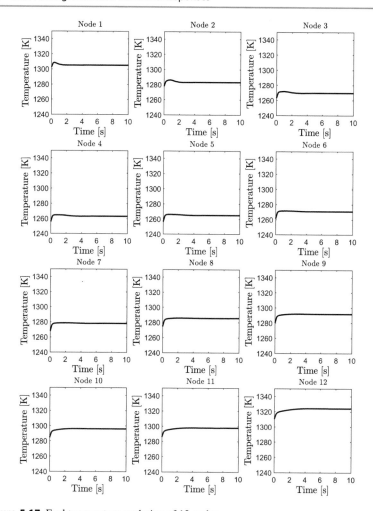

Figure 5.17 Fuel temperature evolution of 12 nodes

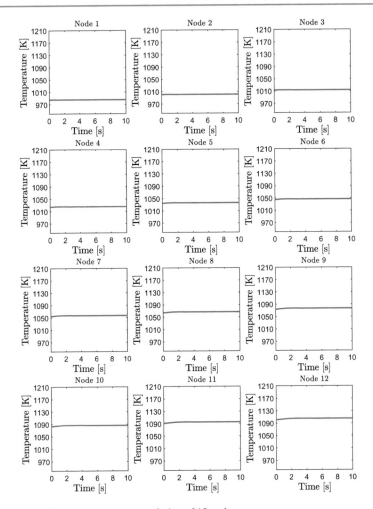

Figure 5.18 Coolant temperature evolution of 12 nodes

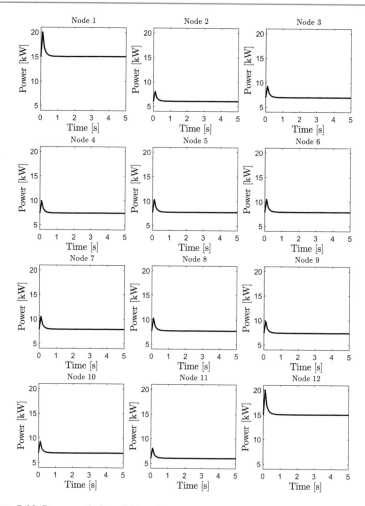

Figure 5.19 Power evolution of 12 nodes

circulation. Since the upstream liquid fuel in node 1 to node 6 is already heated by the increased power immediately after the transient and then sent to the downstream nodes, a slight increase in fuel temperature occurs in the downstream nodes despite the decreased power. It should be noted that for the SMDFR, the transient responses of the global reactivity insertion accident are different from those of conventional reactors in which the fuel is solid, and the temperature evolution of the entire fuel has a similar shape to that in node 1 to node 6 of the SMDFR.

Due to the large mass flow rate, the coolant temperature has only a minimal increase during the transient, which means that the heat absorbed by the coolant before and after the transient is almost unchanged.

As shown in Figure 5.20, the profile of the fuel temperature distribution before and after the transient is similar with a slight increase in magnitude, while the coolant temperature distribution remains unchanged. From Figure 5.21 and Figure 5.22, it can be observed that the fission power after the transient increases by a factor of 27% in less than one second and then decreases to a constant value of 102 kW due to the negative temperature feedback. Finally, the system reaches a new steady state and operates at 105% of its nominal power. The profiles of the DNP concentrations are not changed during the transient, while the magnitude has a slight increase due to the increased neutron flux, as shown in Figure 5.23 (Figure 5.18).

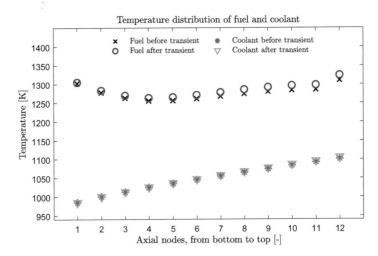

Figure 5.20 Temperature distribution of fuel and coolant

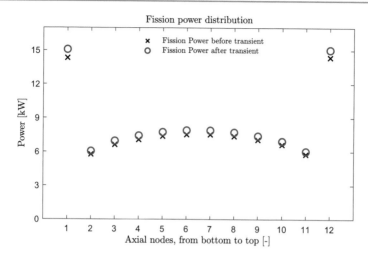

Figure 5.21 Fission power distribution before/after transient

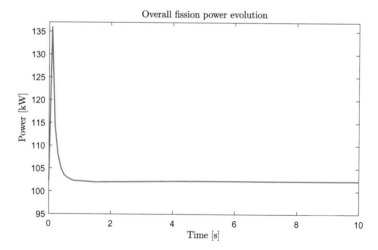

Figure 5.22 Overall fission power evolution

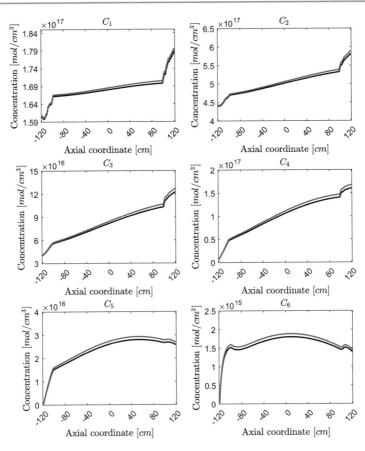

Figure 5.23 DNP concentrations inside the core before (black) and after (red) the transient: reactivity insertion accident

5.4.2   SMDFR Single Channel Transient: Decrease of Fuel Inlet Temperature

In this section, the fuel inlet temperature was decreased by 30 K due to a certain malfunction of the online fuel processing procedure, and the corresponding system responses were investigated.

In this fuel inlet temperature decrease accident, the temperature and power evolution of the 12 axial nodes are shown in Figure 5.24 to Figure 5.26. For node 1 to

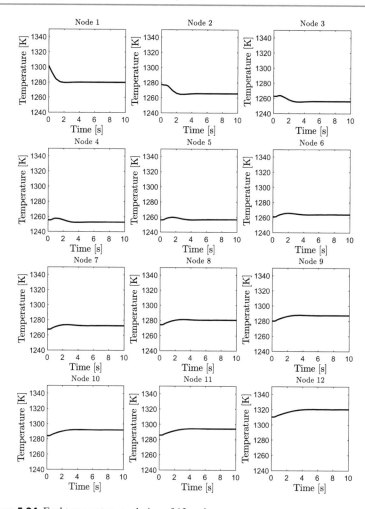

Figure 5.24 Fuel temperature evolution of 12 nodes

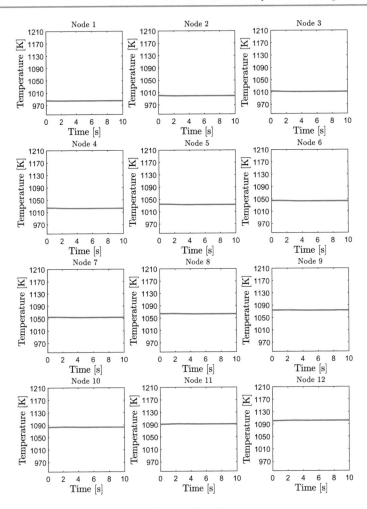

Figure 5.25 Coolant temperature evolution of 12 nodes

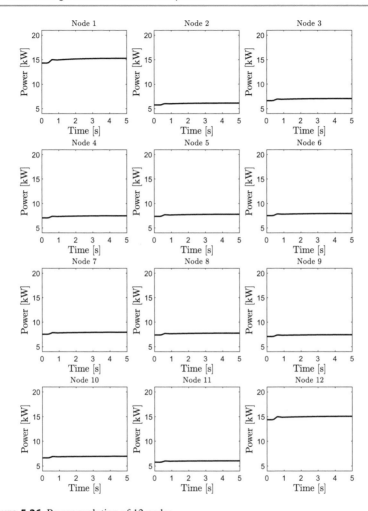

Figure 5.26 Power evolution of 12 nodes

node 4, the fuel temperature decreases as the inlet temperature is decreases. However, as a positive reactivity is introduced by the decreased fuel temperature in these four nodes, the fission power is increased so that the fuel temperatures in the remaining nodes are even higher than the original values. Due to the large heat capacity rate, the coolant temperature has only a minimal increase during the transient, and the heat absorbed by the coolant before and after the transient is almost unchanged.

As shown in Figure 5.27, the profile of the fuel temperature distribution after the transient is slightly changed, with lower values in the inlet region and higher values in the outlet region. No difference is observed for the profile of the coolant temperature distribution before and after the transient.

The fission power distribution and evolution are shown in Figure 5.28 and Figure 5.29. After the transient the power distribution has an unchanged shape with a higher magnitude. The fission power gradually increases by a factor of 6% in four seconds and then remains at a constant value of 103 kW. Finally, the system reaches a new steady state and operates at 106% of its rated power. This means that even if the fuel inlet temperature is reduced by 30 K and no control system is activated, the reactor is able to stabilize itself at a higher power level with the coolant outlet temperature unchanged. The profiles of the DNP concentrations are not changed during the transient, while the magnitude has a slight increase due to the increased neutron flux, as shown in Figure 5.30 (Figure 5.25).

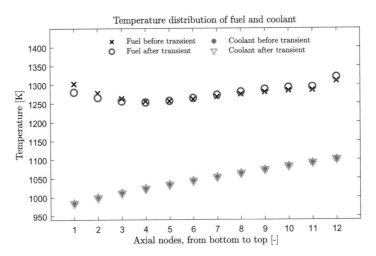

Figure 5.27 Temperature distribution of fuel and coolant

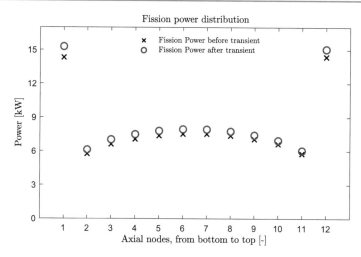

Figure 5.28 Fission power distribution before/after transient

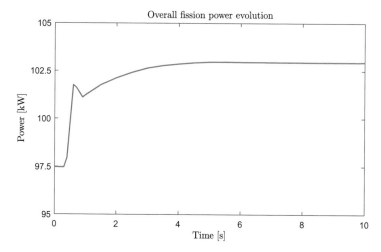

Figure 5.29 Overall fission power evolution

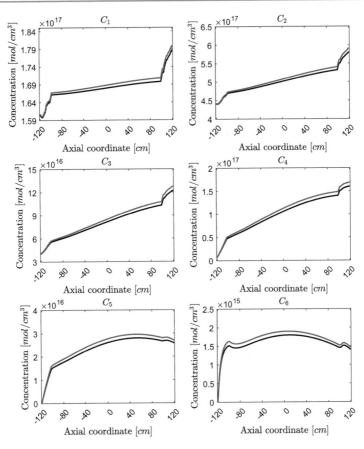

Figure 5.30 DNP concentrations inside the core before (black) and after (red) the transient: decrease of fuel inlet temperature

5.4.3   SMDFR Single Channel Transient: Decrease of Fuel Mass Flow Rate

In this section, the fuel mass flow rate was reduced to 50% of its nominal value due to a specific accident, and the corresponding system responses were investigated.

In this fuel mass flow rate reduction accident, the temperature and power evolution of the 12 axial nodes are shown in Figure 5.31 to Figure 5.33. For node 1 and node 12, the heat transfer between fuel and coolant decreases due to the reduced

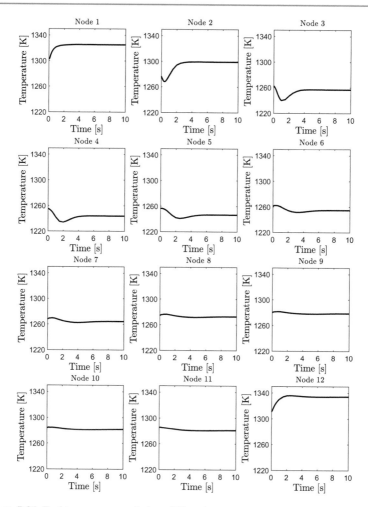

Figure 5.31 Fuel temperature evolution of 12 nodes

mass flow rate, and thus the fuel temperature becomes higher because less heat is absorbed by the coolant. For node 2 to node 11, the temperature initially drops and then rises or becomes constant in a short time (within 3 s). Initially, as the power is reduced due to the temperature feedback in node 1 and node 12, less heat is absorbed by the fuel and thus its temperature is reduced. As the temperature in the nodes of

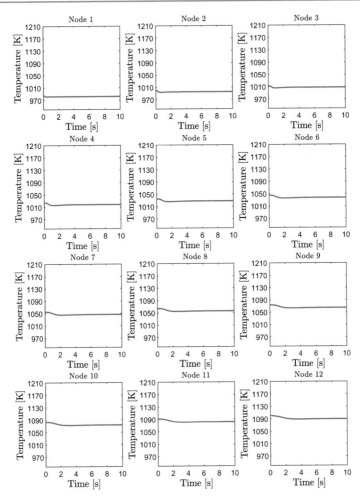

Figure 5.32 Coolant temperature evolution of 12 nodes

the core zone decreases, positive reactivity is introduced due to the temperature feedback effect, resulting in increased power. At the same time, the hot fluid from node 1 passes through these nodes and contributes to the temperature increase. It should be noted that the temperature evolution of these 12 nodes are not only determined by the deposited heat, but also by the convection from the upstream fluid.

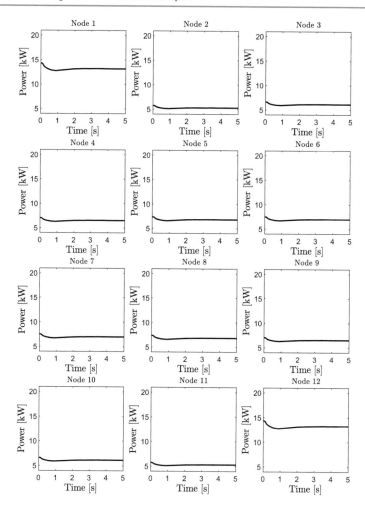

Figure 5.33 Power evolution of 12 nodes

Due to the large heat capacity rate, the coolant temperature has a slight decrease (within 10 K) during the transient, and the heat absorbed by the coolant before and after the transient is slightly reduced.

As shown in Figure 5.34, the profile of fuel temperature distribution after the transient is slightly changed, with higher values at the inlet and outlet regions. The profile of the coolant temperature distribution after the transient is similar, but has a slightly lower magnitude.

The fission power distribution and evolution are shown in Figure 5.35 and Figure 5.36. After the transient the power distribution has an unchanged shape with a lower magnitude. The fission power decreases by a factor of 11% in one second and then increases to a constant value of 90 kW due to the negative temperature feedback effect. Finally, the system reaches a new steady state and operates at 92% of its rated power. This means that if the fuel inlet mass flow rate is reduced by 50% and no control system is activated, the reactor is able to stabilize itself with a lower fission power (92%) and a slightly lower coolant outlet temperature (-10 K).

Figure 5.37 shows that the profiles of the DNP concentrations were significantly affected by this transient, since the convection term plays an important role in the DNP transport process and is determined by the fuel flow field. In addition, the DNP concentrations are also influenced by the fission production rate and the decay rate, which are an effect of competition between each other. This drift effect has a

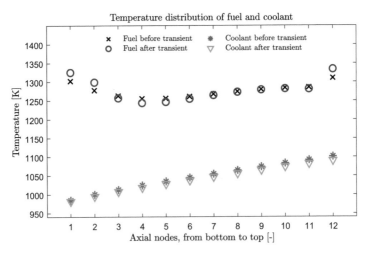

Figure 5.34 Temperature distribution of fuel and coolant

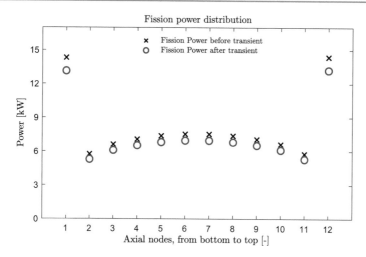

Figure 5.35 Fission power distribution before/after transient

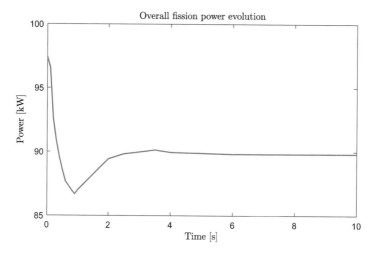

Figure 5.36 Overall fission power evolution

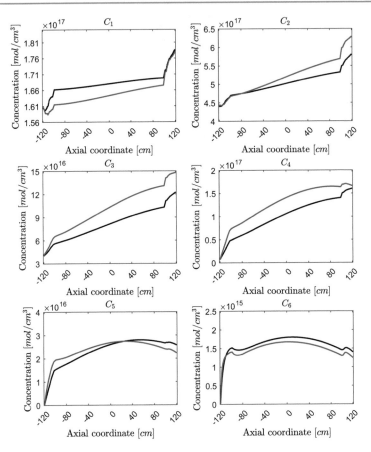

Figure 5.37 DNP concentrations inside the core before (black) and after (red) the transient: decrease of fuel mass flow rate

certain influence on the transient behavior through its contribution to the neutron flux, especially when the fuel mass flow rate is changed significantly, so that special attention has to be paid to address this effect in the safety analysis (Figure 5.32).

5.4.4   SMDFR Single Channel Transient: Decrease of Coolant Inlet Temperature

In this section, the coolant inlet temperature was decreased by 30 K due to a specific accident and the corresponding system responses were investigated.

In this coolant inlet temperature reduction accident, the temperature and power evolutions of the 12 axial nodes are shown in Figure 5.38 to Figure 5.40. The fuel

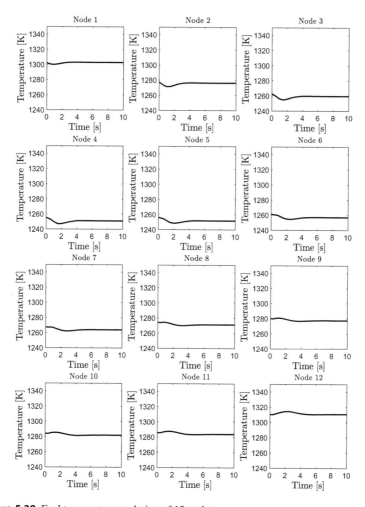

Figure 5.38 Fuel temperature evolution of 12 nodes

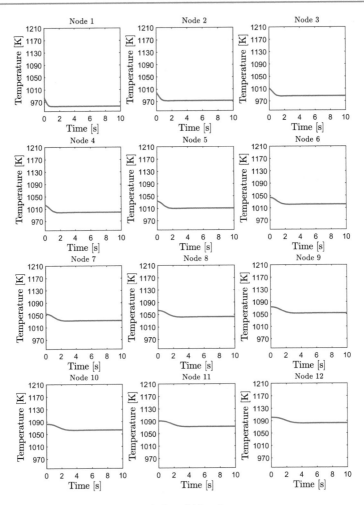

Figure 5.39 Coolant temperature evolution of 12 nodes

temperature is initially decreases (within 3 seconds), and then increases to a new value that is close to its original value. As the coolant inlet temperature is lowered, the fuel temperature is lowered as well, because the heat transfer between them is enhanced due to the larger temperature difference, and more heat is absorbed by the coolant. At the same time, the power increases due to the negative temperature

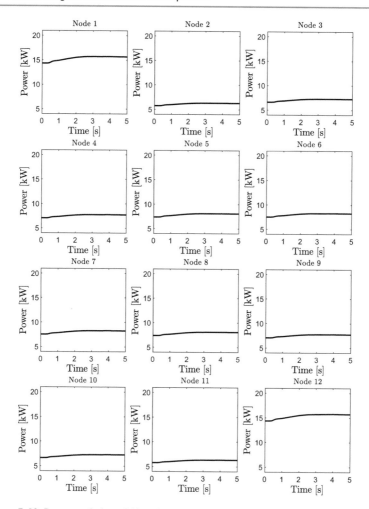

Figure 5.40 Power evolution of 12 nodes

feedback effect. As the power increases, negative reactivity is introduced into the system due to the increased temperature. Eventually, a new steady state is reached. The coolant temperatures of all the 12 nodes are reduced by approximately 30 K as the coolant inlet temperature is reduced by this value.

As shown in Figure 5.41, the profile of the fuel temperature distribution before and after the transient is almost the same. The profile of the coolant temperature

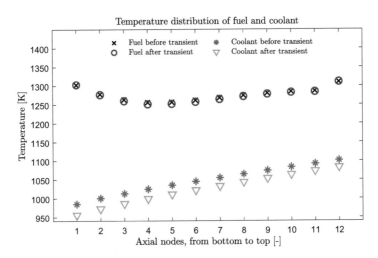

Figure 5.41 Temperature distribution of fuel and coolant

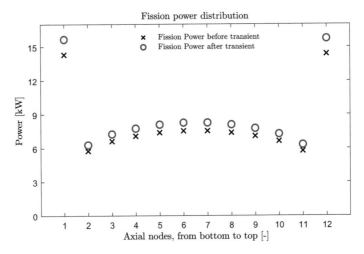

Figure 5.42 Fission power distribution before/after transient

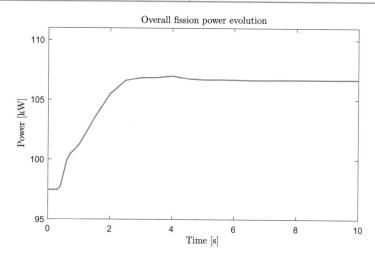

Figure 5.43 Overall fission power evolution

distribution before and after the transient is similar, but has a smaller magnitude difference of about −25 K.

The fission power distribution and evolution are shown in Figure 5.42 and Figure 5.43. After the transient the power distribution has an unchanged shape with a higher magnitude. The fission power gradually increases by a factor of 10% in three seconds and then decreases slightly to a constant value of 107 kW. Finally, the system reaches a new steady state and operates at 110% of its rated power. This means that even if the coolant inlet temperature is reduced by 30 K and no control system is activated, the reactor is able to stabilize itself with a higher fission power and a lower coolant outlet temperature ($-20\ K$). The DNP concentrations do not change, except for a slight magnitude increase during the transient, as shown in Figure 5.23 (Figure 5.39) (Figure 5.44).

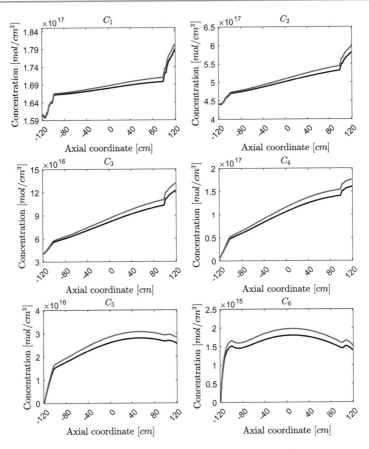

Figure 5.44 DNP concentrations inside the core before (black) and after (red) the transient: decrease of coolant inlet temperature

5.4.5 SMDFR Single Channel Transient: Decrease of Coolant Mass Flow Rate

In this section, the coolant mass flow rate was reduced to 50% of its nominal value due to a specific accident and the corresponding system responses were investigated.

In this coolant mass flow rate reduction accident, the temperature and power evolutions of the 12 axial nodes are shown in Figure 5.45 to Figure 5.47. Initially, the fuel temperature increases with a delay of about 0.5 s, as the coolant temperature is

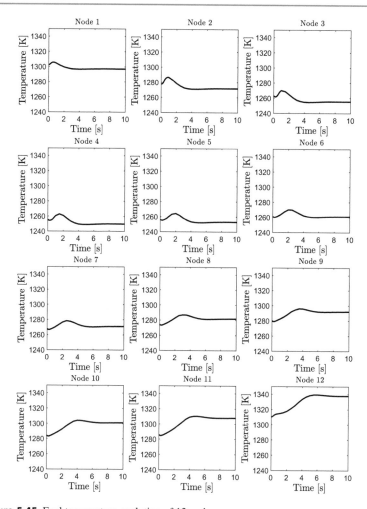

Figure 5.45 Fuel temperature evolution of 12 nodes

increases due to the deterioration of the heat transfer and the reduced heat capacity rate of the coolant. As the power decreases due to the negative temperature feedback effect, the fuel temperature stops increasing and then decreases to a constant value. The power is initially increased by the positive feedback effect of the coolant temperature, and then decreases as the fuel temperature increases.

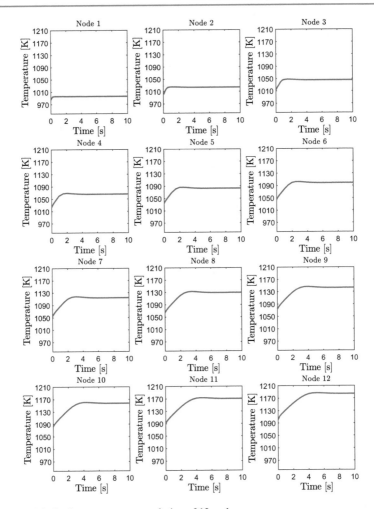

Figure 5.46 Coolant temperature evolution of 12 nodes

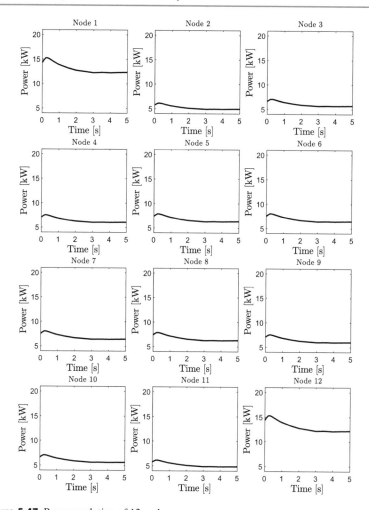

Figure 5.47 Power evolution of 12 nodes

As shown in Figure 5.48, the profile of the fuel temperature distribution after transient is slightly changed, with higher values at the exit regions. The coolant temperature deviation after the transient increases from bottom to top because the heat capacity rate decreases and more heat is absorbed by the same amount of coolant.

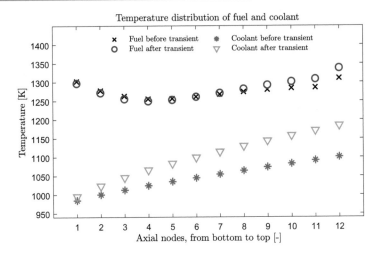

Figure 5.48 Temperature distribution of fuel and coolant

The fission power distribution and evolution are shown in Figure 5.49 and Figure 5.50. After the transient the power distribution has an unchanged shape with

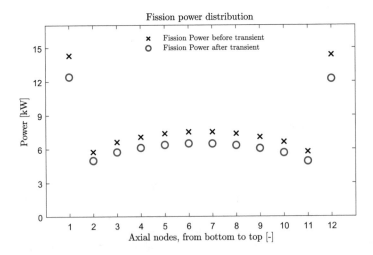

Figure 5.49 Fission power distribution before/after transient

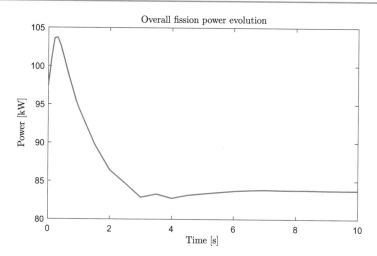

Figure 5.50 Overall fission power evolution

a lower magnitude. The fission power increases by a factor of 6% in 0.5 seconds and then decreases to a constant value of 84 kW due to the negative temperature feedback effect. Finally the system reaches a new steady state and operates at 86% of its rated power. This means that when the fuel inlet mass flow rate is reduced by 50% and no control system is activated, the reactor is able to stabilize itself with a lower fission power (86%) and a significantly increased coolant outlet temperature (+90 K). The DNP concentrations do not change, except for a slight magnitude decrease during the transient, as shown in Figure 5.51 (Figure 5.46).

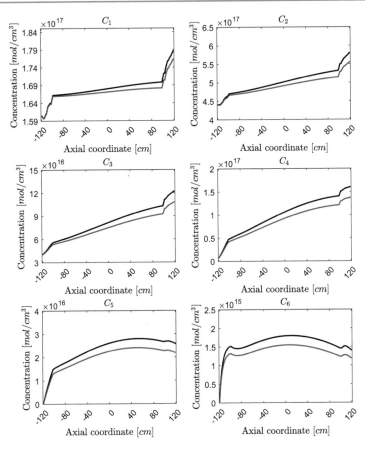

Figure 5.51 DNP concentrations inside the core before (black) and after (red) the transient: decrease of coolant mass flow rate

5.5   Uncertainty and Sensitivity Analysis of SMDFR Single Channel Transient Responses

In this section five perturbation parameters (Table 5.2) are selected and applied to the SMDFR single channel model to investigate the system responses under perturbation and to quantitatively assess the importance level of these five parameters.

Table 5.2 Perturbation parameters

#	Parameters	Input of Code	Min.	Max.
1	Inserted reactivity (value [pcm])	COMSOL	−10	10
2	Fuel inlet temperature deviation (additive [K])	COMSOL	−10	10
3	Fuel velocity deviation (factor [-])	COMSOL	−0.05	0.05
4	Coolant inlet temperature deviation (additive [K])	COMSOL	−10	10
5	Coolant velocity deviation (factor [-])	COMSOL	−0.05	0.05

5.5.1   Uncertainty Analysis of SMDFR Single Channel Transient Responses

In order to quantify the uncertainties of simulation results, an error bar was used to illustrate the two-sided tolerance limits: the length below the data point was determined by the lower limit and the length above the data point dipicted the upper limit.

As shown in Figure 5.52, the uncertainty band of the average temperature of fuel is quite narrow, which is within ±10 K. The peak temperature is depicted in Figure 5.53, where the uncertainty of the fuel peak temperature is also kept within a minimum range and an upper limit of +20 K is observed for the peak temperature of the coolant in the core zone. Since the maximum uncertainty of the average and peak temperature of fuel and coolant is not beyond ±20 K, the operational stability of the SMDFR system under disturbance is verified.

Referring Figure 5.54, a relative high uncertainty band (power deviation of around ±1 kW) under disturbance is observed in node 1 and node 12. For the rest nodes, the power deviation is within ±0.5 kW. The relative overall power deviation is not more than 6% as shown in Figure 5.55. Even a minimum deviation of fuel and coolant temperature is observed, the thermal power has to be regulated by the control system so that the desired/nominal power can be maintained under disturbance.

As shown in Figure 5.56, the largest heat taken by coolant occurs in node 1 (±1.5 kW) due to the largest temperature difference between fuel and coolant. For the rest nodes, the uncertainty of transferred heat is within ±0.5 kW.

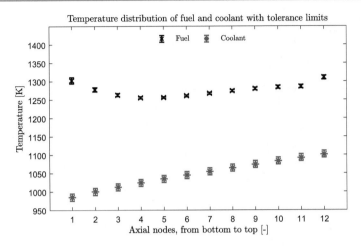

Figure 5.52 Temperature distribution of fuel and coolant with tolerance limits in a perturbed state

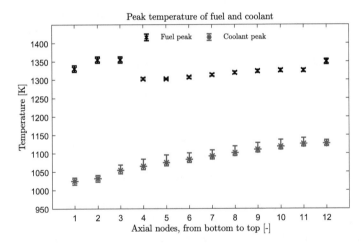

Figure 5.53 Peak temperature of fuel and coolant with tolerance limits in a perturbed state

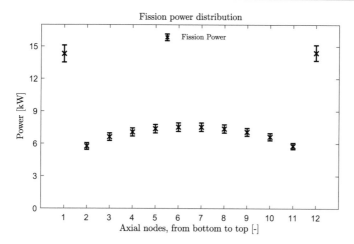

Figure 5.54 Fission power distribution with tolerance limits in a perturbed state

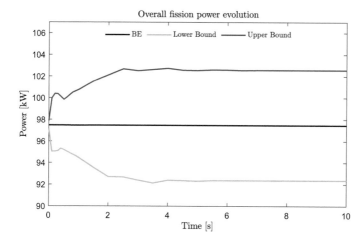

Figure 5.55 Overall fission power evolution with tolerance limits under disturbance

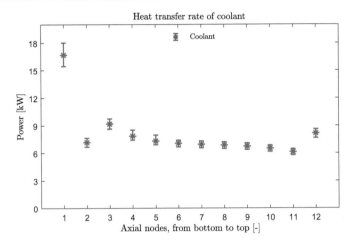

Figure 5.56 Heat transfer rate of coolant with tolerance limits in a perturbed state

5.5.2   Sensitivity Analysis of SMDFR Single Channel Transient Responses

In order to quantify the influence of disturbances on the transient responses, the total power is selected as the output variable to be analyzed. By performing the sensitivity analysis based on the regression-based technique and the variance-based technique separately, the importance level of five perturbation parameters (Table 5.2) on the transient responses are quantitatively assessed.

As shown in Figure 5.57, the coefficient of multiple determination is 0.9953 (almost equal to 1.0), which means that a linear correlation is considered as the dominant part when evaluating the influence of perturbation parameters on the power. As shown in Figure 5.58, there is almost no discrepancy between the significance levels obtained by the regression-based method, by the first-order index and by the total-effect index. Parameter 4, the coolant inlet temperature deviation, has the largest influence on the output variable. This means that it has the highest priority to keep the coolant inlet temperature constant during normal operation in order to make the undesired power deviation as low as possible. In addition, Parameter 2, the fuel inlet temperature deviation, has the second most important influence on the system output variable and should also be kept constant. The remaining parameters have almost no influence on the system output and can be considered as negligible on the power. This means that this system has a high resistance to the reactivity

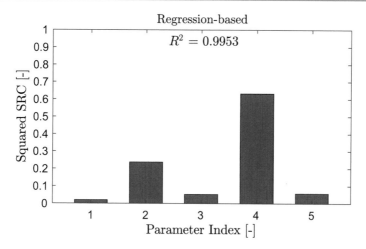

Figure 5.57 Parameter sensitivity to post-transient power using regression-based method

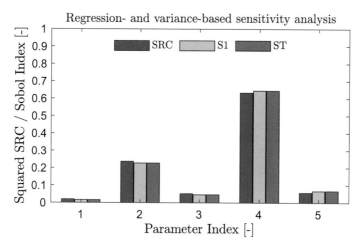

Figure 5.58 Parameter sensitivity to post-transient power using regression- and variance-based method

disturbance due to the strong negative feedback effect. In addition, the small deviation ($\pm 5\%$) of the mass flow rates of two fluids has no effect on the thermal power, which means that the system is resistant to the flow fluctuation and can maintain the power unchanged under these disturbances.

Results and Analysis: SMDFR Full Core

6

This chapter is divided into three parts: the results of the SMDFR full core model, the uncertainty-based system model of the SMDFR core, and the control system of the SMDFR core and its optimization.

6.1 SMDFR Full Core Steady State

The temperature and power distribution of fuel and coolant of the SMDFR core at steady state are shown in Figure 6.1 and Figure 6.2, respectively. For the temperature distribution there is an obvious hot region in the upper left corner, where the fuel is heated up to 1400 K. Since the fuel there is not as well circulated as in the other regions, a deterioration of the heat transfer is to be expected. Also, according to Figure 6.2, the power density in the center area is higher than that in the periphery, since the maximum neutron flux exists on the center line. In addition, although the power density in the distribution and collection zones is lower than that in the core zones, the total power in the distribution and collection zones is higher than that in the core zones due to a much higher fuel content with a factor of about 2.5.

C. Liu, *Multiscale and Multiphysics Modeling of Nuclear Facilities with Coupled Codes and its Uncertainty Quantification and Sensitivity Analysis*, https://doi.org/10.1007/978-3-658-43422-9_6

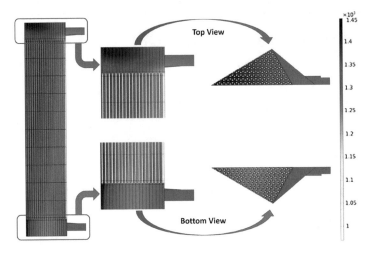

Figure 6.1 Temperature distribution of fuel and coolant of the SMDFR core (color legend unit in $[K]$)

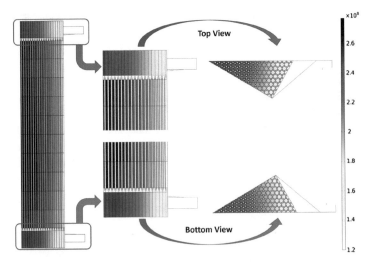

Figure 6.2 Power density distribution of the SMDFR core (color legend unit in $[W/m^3]$)

The fuel flow is shown by the velocity streamline and speed arrow in Figure 6.3 and Figure 6.4, respectively. Starting from the inlet nozzle, the fuel is pumped into the distribution zone and then distributed in the gap between the coolant pipes. After that, the flow direction is changed to be parallel to the tube bundle and then the fuel passes through the core zones. In the collection zone, the fuel flow direction is changed again to be parallel to the outlet nozzle, and then it leaves the core and passes through the online fuel processing unit for the next round of circulation. It is obvious that there is a strong interaction between the fuel and the wall at the top of the core, and a special attention has to be paid to this area because the risk of structural material failure was relatively high and the vibration could occur here.

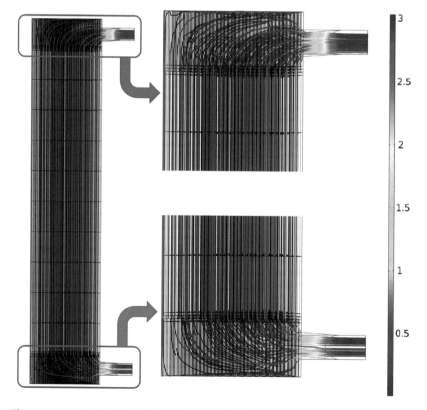

Figure 6.3 Velocity streamline of the fuel of the SMDFR core (color legend unit in [m/s])

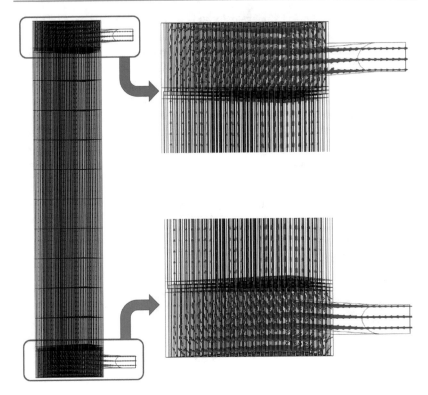

Figure 6.4 Speed arrow of fuel of the SMDFR core (in blue)

The flow of the coolant is shown by the speed arrow in Figure 6.5. No change in velocity direction is observed, and the velocity is always parallel to the tube bundle from bottom to top. Based on this, it is reasonable to assume in the system model that the coolant velocity is in the axial direction, with no radial component.

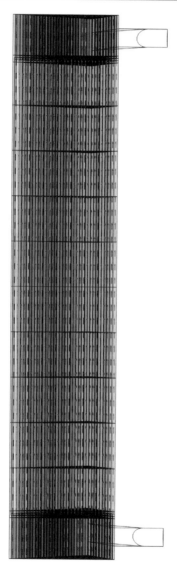

Figure 6.5 Speed arrow of coolant of the SMDFR core (in red)

6.2   Uncertainty-based System Model of the SMDFR Full Core

Since the SMDFR is a new concept, and no experimental data are available at this stage, the 1D Matlab model has to be verified against the high-fidelity SMDFR core model, and the uncertainty of its parameters has been quantified using the inverse uncertainty propagation technique to compensate for the errors introduced by the 1D system model (Table 6.1). A comparison is shown in Figure 6.6, where the asterisk represents the fuel temperature and the cross denotes the coolant temperature. The results from the high-fidelity model are shown in red and those from the Matlab model are shown in black. Using the high-fidelity model results as a reference, it can be seen that the shapes of the fuel and coolant temperature distributions are generally well captured by the Matlab model. Except for Node 12, the temperature differences at all other nodes are within ±20°C. In addition, the tolerance limits are able to cover all the data provided by the high-fidelity model, which means that the Matlab model has sufficient capacity to capture the system behavior and correctly

Table 6.1 Uncertain parameters of the MATLAB model

#	Parameters	PDF	Min.	Max.
1	Thermal power (factor)	Uniform	0.98	1.02
2	Temperature feedback coefficient of fuel (factor)	Uniform	0.9	1.1
3	Temperature feedback coefficient of coolant (factor)	Uniform	0.9	1.1
4	Mass flow rate of fuel (factor)	Uniform	0.95	1.05
5	Mass flow rate of coolant (factor)	Uniform	0.95	1.05
6	Heat capacity of fuel (factor)	Uniform	0.95	1.05
7	Heat capacity of coolant (factor)	Uniform	0.95	1.05
8	HTC between fuel and wall (factor, Node 1)	Uniform	0.81271	1.5812
9	HTC between fuel and wall (factor, Node 2–11)	Uniform	0.82613	1.0219
10	HTC between fuel and wall (factor, Node 12)	Uniform	0.45038	1.0213
11	HTC between wall and coolant (factor, Node 1)	Uniform	0.9	1.1
12	HTC between wall and coolant (factor, Node 2–11)	Uniform	0.9	1.1
13	HTC between wall and coolant (factor, Node 12)	Uniform	0.9	1.1
14	Fuel inlet temperature (additive)	Uniform	−2.2	2.2
15	Coolant inlet temperature (additive)	Uniform	−2.2	2.2

predict the temperature distributions. Finally, it is verified that the 1D Matlab model with tolerance limits can be used to study the core behavior and to design and optimize the control system.

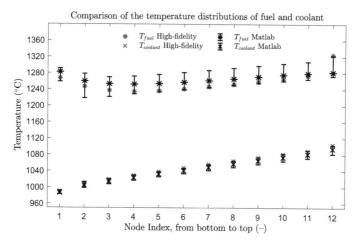

Figure 6.6 Comparison of the temperature distributions of fuel and coolant

6.3   Control System of the SMDFR Core and its Optimization

To investigate the feasibility of the control system, three transients are calculated with and without the control system: ± 100 pcm reactivity insertion, $\pm 20\,^{\circ}$C variation in the inlet temperature, and $\pm 10\%$ variation in the coolant mass flow rate.

The results of the system response for ± 100 pcm reactivity insertion are shown in Figure 6.7 and Figure 6.8. The power evolution of the system without a control system is shown in red and that with a control system is shown in black. Without a control system, as shown in Figure 6.7, a reactivity of 100 pcm is inserted into the core at t = 10 s, causing the power to increase by about 6%. As more power is generated to heat the fuel and coolant, their temperatures are increased. Due to the negative reactivity feedback of the fuel and coolant, a negative reactivity is introduced, and, thus, the power returns to 102.5% of the nominal power and remains constant at this level. However, the power level changes due to this perturbation, which is undesirable during the normal operation when an unchanged power level is expected despite the perturbation. With the control system, the power returns to its nominal value after several oscillations, and the peak value of the power is 1% lower than that without a control system. For the −100 pcm reactivity insertion transient (Figure 6.8), the system with a control system is also able to maintain the nominal

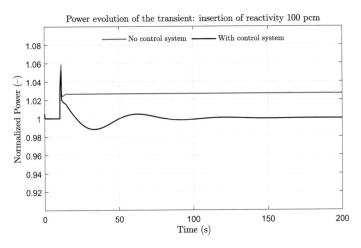

Figure 6.7 System response to the insertion of reactivity: 100 pcm

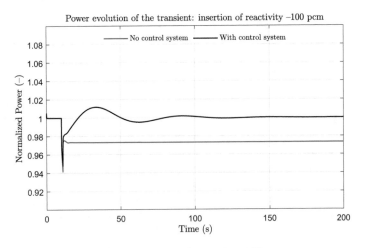

Figure 6.8 System response to the insertion of reactivity: −100 pcm

power after the perturbation. In both cases, the time required for the system to return to its original state is within 150 s.

The results of the system response for ±20 °C variation in the inlet temperature are shown in Figure 6.9 and Figure 6.10. Without a control system, as shown in Figure 6.9, the coolant inlet temperature increases by 20 °C at t = 10 s, causing the power to decrease by about 8% due to the negative reactivity feedback of the coolant. The decrease in power results in a decrease in fuel temperature. Thus, the negative reactivity introduced by the coolant is offset by the positive reactivity introduced by the decrease in fuel temperature. Finally, the power remains constant at the level of 92.5% of its nominal power. With a control system, the power returns to its nominal value after several oscillations, and the peak value of the power is 1% higher than that without a control system. For the transient with the decrease of 20 °C (Figure 6.10), the system with a control system is able to maintain the nominal power after the perturbation. In both cases, the time required for the system to return to its original state is 150 s.

The results of the system response for ±10% variation in the coolant mass flow rate are shown in Figure 6.11 and Figure 6.12. Without a control system, as shown in Figure 6.11, the coolant mass flow rate increases by 10% at t = 10 s, causing the power to increase by about 3.5%. As soon as the coolant mass flow rate increases, the heat transfer between the fuel and the coolant increases and their temperatures

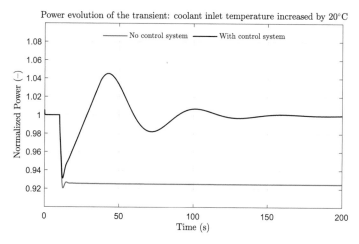

Figure 6.9 System response to the temperature variation: inlet temperature increased by 20 °C

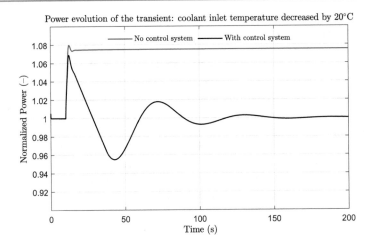

Figure 6.10 System response to the temperature variation: inlet temperature decreased by 20 °C

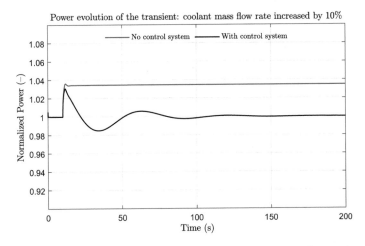

Figure 6.11 System response to the mass flow rate variation: coolant mass flow rate increased by 10%

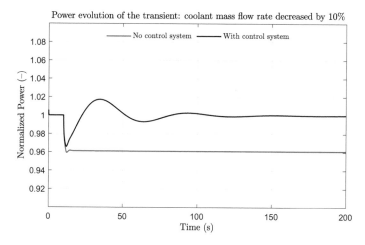

Figure 6.12 System response to the mass flow rate variation: coolant mass flow rate decreased by 10%

thus decrease, resulting in the introduction of positive reactivity due to the negative reactivity feedback. As the power increases, the fuel and coolant temperatures increase, and then the positive reactivity introduced is offset by the negative reactivity introduced due to the increase in fuel and coolant temperatures. Finally, the power remains constant at the level of 103.5% of its nominal power. With a control system, the power returns to its nominal value after several oscillations, and the peak value of the power is 1% lower than that without a control system. For the transient with the decrease of the coolant mass flow rate (Figure 6.12), the system with a control system is also able to maintain the nominal power after the perturbation. In both cases, the time required for the system to return to its original state is 150 s.

6.3.1   Optimization

Although the control system is proven to be able to handle various transients, its parameters need to be optimized to achieve the best performance. To optimize of the control system, two scenarios are selected as benchmark cases: step load change: 100% FP to 90% FP; linear load change: 100% FP to 50% FP to 100% FP, as shown in Figure 6.13 and Figure 6.14.

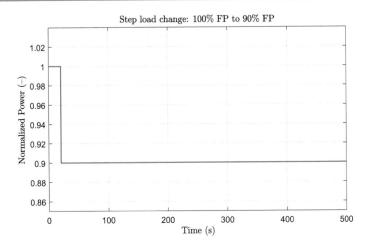

Figure 6.13 Benchmark case: step load change: 100% FP to 90% FP

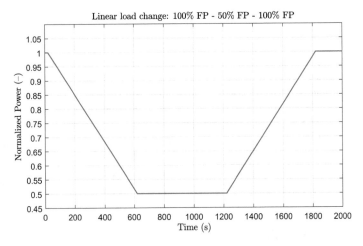

Figure 6.14 Benchmark case: linear load change: 100% FP to 50% FP to 100% FP

The integral time-weighted absolute error (ITAE), as defined by Equation (6.1), is chosen as the control performance criterion for the case of step load change, since the errors that exist after a long time have to be weighted much more heavily than those at the beginning of the transient. For the linear load change, the integral absolute error (IAE), as defined by Equation (6.2), is chosen as the performance criterion because the error in this case should not be time-weighted. In both cases, the uncertain parameters in Table 6.1 are used to calculate the one-sided upper tolerance limit of the ITAE or IAE, which is the output value of the fitness function of the particle swarm optimization (PSO) method. The obtained optimized solution has the following meaning: under the given uncertainty parameter ranges, at least 95% of the possible system states would have a better performance than that of the final optimized value obtained by the PSO method with a confidence level of 95%. By using the optimized parameters, the control system has a confidence level of 95% that it will deliver optimized performance with 95% probability.

$$ITAE = \int_0^{t_{max}} t\,|e(t)|\,dt, \tag{6.1}$$

$$IAE = \int_0^{t_{max}} |e(t)|\,dt, \tag{6.2}$$

where $e(t)$ is the error between the measured value of the controlled variable and its desired value.

Step Load Change

The main parameters for the PSO process are shown in Table 6.2. After 327 generations, the average cumulative change in the value of the fitness function over 50 generations is less than 1×10^{-6}, and the final best point is: Kp = 3944.1, Ki = 494.3, with a fitness value of 55.97. The evolution of the fitness value during the PSO process is shown in Figure 6.15, where the maximum and minimum values

Table 6.2 Major parameters for the PSO process of step load change

Parameters	Value
Fitness function	$\int_0^{t_{max}} t\,|e(t)|\,dt$
Number of variables	2
Constraints	none
Kp	10~10,000
Ki	10~1000
Population size	20
Generations	500
Social attraction	1.25

of the particles in each generation are represented by the error bar. For the first 70 generations, large variations can be observed, as the particles went through the entire domain to identify the best point (Figure 6.16 and Figure 6.17). Then, they started to focus on a smaller region, where the best point can be found. After 180

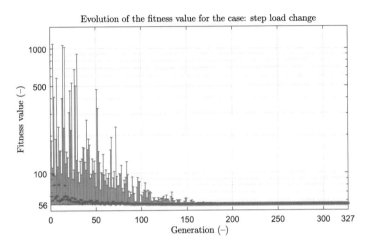

Figure 6.15 Evolution of the fitness value for the case of step load change

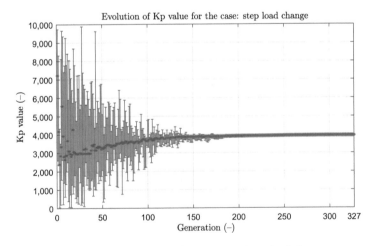

Figure 6.16 Evolution of the parameter Kp for the case of step load change

generations, all particles were close to the final best point, and the optimization process was finished after 327 generations.

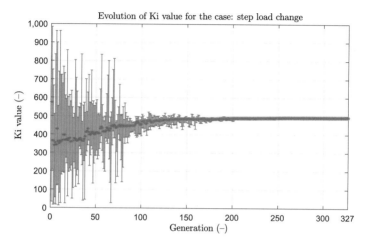

Figure 6.17 Evolution of the parameter Ki for the case of step load change

Linear Load Change

The main parameters for the PSO process are kept the same as those for the step load change, as shown in Table 6.2. After 60 generations, the average cumulative change in the value of the fitness function over 50 generations is less than 1×10^{-6}, and the final best point is: Kp = 10,000.0, Ki = 1000.0, with a fitness value of 0.0127. The evolution of the fitness value during the PSO process is shown in Figure 6.18, where the maximum and minimum values of the generation are represented by the error bar. In contrast to the step load change, the particles found the location of the best point after only 20 generations (Figure 6.19 and Figure 6.20), and the optimization process was terminated after 60 generations. The number of generations needed to identify the best point is much lower because it is not so challenging for the control system to follow the load for the linear load change. Without the preset limits, Kp and Ki would have different values because they have already reached their limits. However, the integral absolute error considering the limits is less than 0.013, which means that a well-optimized control system is obtained and no further optimization is needed.

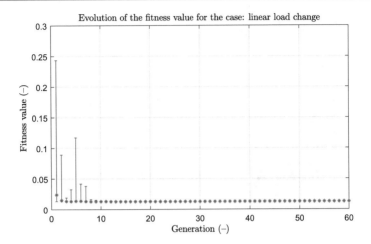

Figure 6.18 Evolution of the fitness value for the case of linear load change

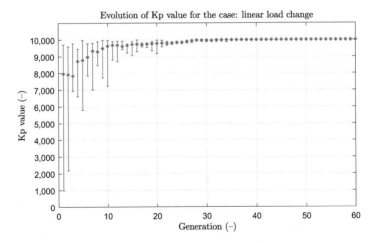

Figure 6.19 Evolution of the parameter Kp for the case of linear load change

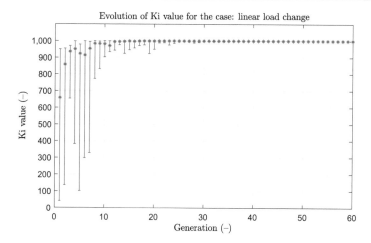

Figure 6.20 Evolution of the parameter Ki for the case of linear load change

6.3.2 Performance Assessment

The performance of the optimized control system must be evaluated considering its uncertainty and compared to the performance before optimization. Unlike the tolerance limits technique used in the optimization process, a Monte-Carlo method is used to evaluate the performance before and after optimization: 1000 uncertain runs are performed, and their confidence intervals (95% confidence) for the 95% and 5% quantiles are chosen as the upper and lower boundaries, respectively.

For the step load change, the reactivity introduced by the control system during the transient is shown in Figure 6.21, and the system responses before and after optimization are shown in Figure 6.22 and Figure 6.23. The desired value is shown in red; the median value of the real power (system response) is shown in black; the blue and green lines represent the upper (95%) and lower (5%) uncertain limits of the system response, respectively.

The evolution of the introduced reactivity after the optimization can be seen in Figure 6.21. In the first stage of the transient (from t = 10 s to t = 60 s), the introduced reactivity reached its lower limit (−15 pcm/s), as the PI controller tried to eliminate the error introduced by the step signal as much as possible and thus made the output value equal to its lower limit in order to reduce the power. In the second stage (from t = 60 s to t = 90 s), since the reactor power was reduced by the introduced negative reactivity introduced in the first stage, the fuel and coolant temperatures were lower than their original values. Due to the negative reactivity feedback coefficient of

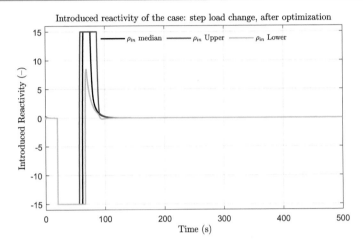

Figure 6.21 Introduced reactivity of the case of step load change

both fluids (Table 2.1), an additional positive reactivity was introduced and thus the power was higher than the set value. To counteract this effect, the PI controller switched its output from the lower limit to the upper limit (15 pcm/s) to reduce the reactor power. Eventually, the power reached its set value at a new steady state, and then the output signal of the PI controller returned to zero.

It is clear that the system response cannot be represented by a single fixed curve in time, but by a region that is bounded by the upper and lower uncertain limits, which means that the system performance cannot be judged by a simple curve, but can be obtained from the results of statistics with a certain degree of confidence. After optimization, the performance of the control system is greatly improved, as shown in Figure 6.22 and Figure 6.23, despite the existence of uncertainty, as it has less overshoot, less oscillation, and less time consumed to reach the new steady state.

The integral time-weighted absolute errors before and after optimization are shown in Figure 6.24 and Figure 6.25. Before optimization, the upper limit, the largest value that the system could deliver is about 270 s^2, when taking into account the existence of uncertainty with a confidence of 95%, and this value is reduced to 88.6 s^2 after optimization, which is consistent with the results presented in Section 6.3.1. Considering the uncertainties of the computer model and the physicochemical properties, the integral time-weighted absolute error of the optimized system has a probability of 95% to stay below 89 s^2 with a confidence of 95%.

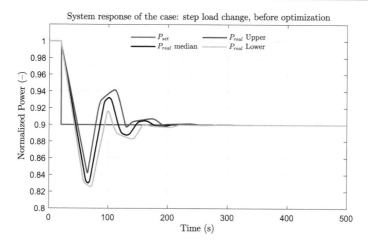

Figure 6.22 System responses to the case of step load change before optimization

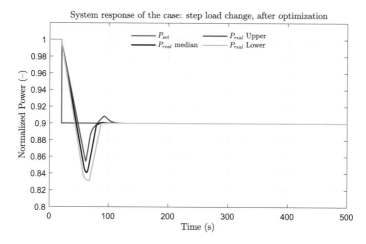

Figure 6.23 System responses to the case of step load change after optimization

The system responses and IAE of the second case: linear load changes are shown in Figures 6.26 to 6.29. It can be seen that the system response matches the desired power very well, and there is almost no discrepancy between the curves, which means that the control system can adjust the power according to the linear load change without any significant deviations or oscillations. However, before opti-

Integral time-weighted absolute error of the case: step load change, before optimization

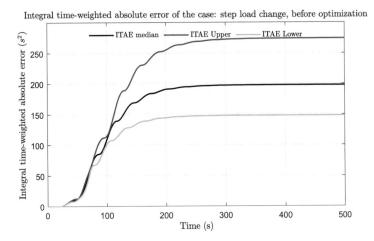

Figure 6.24 Integral time-weighted absolute error of the case of step load change before optimization

Integral time-weighted absolute error of the case: step load change, after optimization

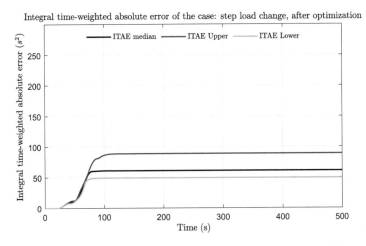

Figure 6.25 Integral time-weighted absolute error of the case of step load change after optimization

mization, small oscillations are observed immediately after the power ramps, which are eliminated after optimization.

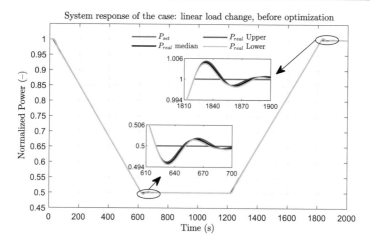

Figure 6.26 System responses to the case of linear load change before optimization

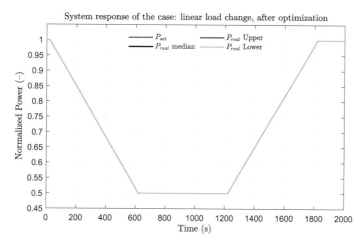

Figure 6.27 System responses to the case of linear load change after optimization

The upper limit of IAE is reduced from 0.7 s to 0.015 s by optimization. Considering the uncertainties, the integral absolute error of the optimized system has a probability of 95% to stay below 0.015 s with a confidence of 95% (Figure 6.27 and 6.28).

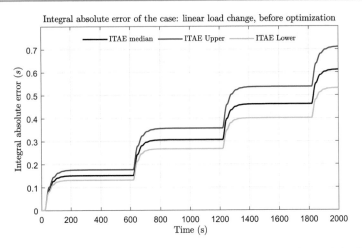

Figure 6.28 Integral absolute error of the case of linear load change before optimization

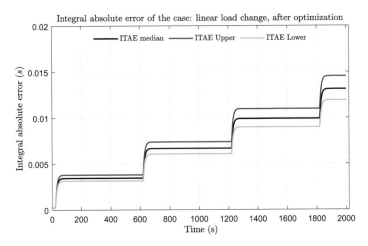

Figure 6.29 Integral absolute error of the case of linear load change after optimization

6.3.3   Sensitivity Analysis

As shown in Figure 6.30, parameter 2, the temperature feedback coefficient of the fuel, has a major influence on the performance of the control system. The reason is that this reactivity feedback plays an important role in adjusting the neutron flux, and

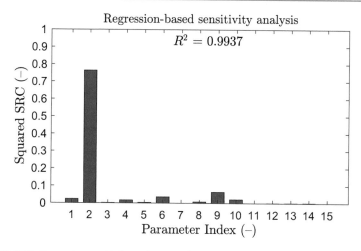

Figure 6.30 Parameter sensitivity to the control system performance using regression-based method

thus the power is changed. Moreover, since its value is several times larger than that of other reactor concepts due to the large expansion rate of molten salt, its impact on power is significant, and determining its value should be prioritized to provide a reliable model with less uncertainty. In addition, parameter 9, the heat transfer coefficient between fuel and wall in the core zone, has a non-negligible influence on the output power. Careful calibration of the heat transfer coefficient has to be performed to reduce the power uncertainty introduced by the heat transfer process. Since the coefficient of multiple determination R^2 is quite close to 1.0, the system response can be assumed to be completely linear. This means that the results provided by the regression-based technique are of the same high quality as those provided by the variance-based technique. However, this does not mean that the system model is linear. This linear relationship can only be applied between the input variables and the selected system response: the performance of the control system. From Figure 6.31 it can be observed that for parameters 2 and 8, the total effect indices, ST, are slightly higher than the first-order sensitivity indices, S1, which means that these two parameters have certain co-effects on the system response together with other parameters, and these effects have been captured by the total-effect indices.

Another way to reduce uncertainty is to use a high-fidelity model, such as the coupled Monte-Carlo and CFD model. However, this will significantly increase the computational cost and is therefore not suitable for the design and optimization

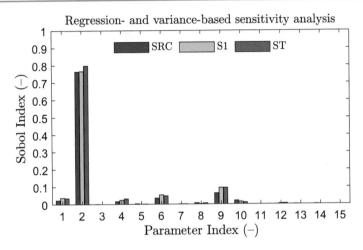

Figure 6.31 Parameter sensitivity to the control system performance using regression- and variance-based method

of the control system under the given computational resources. The Matlab built in this dissertation offers a compromise between model accuracy and its computational cost, which could perfectly cover the introduced error by uncertainty bounds.

Summary, Conclusion and Outlook 7

7.1 Summary

7.1.1 Summary of the Uncertainty Quantification and Sensitivity Analysis Methodology

To quantify the output uncertainty, the concept of tolerance limits was applied. This technique reduced the number of runs to 93, which was acceptable for the expensive coupled computer model. In addition, the inverse uncertainty propagation method was used in the system modeling in MATLAB to account for the loss of precision due to dimensionality reduction.

Four groups of methods were available for the sensitivity analysis: Graphical methods, Screening analysis, Regression-based analysis and Variance-Based analysis. Regression-based analysis is the only feasible method for quantitative sensitivity analysis by using the original computer model when the linear relationship is dominant. By introducing the fast-running surrogate model, a global effect (including the non-linear relationship) can be assessed using variance-based methods. When the model has a linear (or monotonic) response to the uncertain parameters, the regression-based methods can provide a consistent output as the variance-based methods at a much lower computational cost. However, when non-linear (and non-monotonic) and/or interactions are present, variance-based methods are preferred for higher accuracy.

© The Author(s), under exclusive license to Springer Fachmedien Wiesbaden GmbH, part of Springer Nature 2023
C. Liu, *Multiscale and Multiphysics Modeling of Nuclear Facilities with Coupled Codes and its Uncertainty Quantification and Sensitivity Analysis*,
https://doi.org/10.1007/978-3-658-43422-9_7

7.1.2   Summary of the Simulation Results

In this dissertation, the characteristics of the cooling system using liquid metal as the working fluid are investigated in the following steps:

1. A coupled system and 3D computational fluid dynamics model of an experimental loop for both local and global responses: in the framework of the EU-Project SESAME, an experimental facility called TALL-3D loop has been built to provide experimental data for the code validation and assessment in the numerical simulation of liquid metal coolant loops. The benchmark test TG03.S301.03, dedicated to the investigation of the limit cycle oscillations in the primary circuit of the TALL-3D facility, was selected as the case to be investigated in this dissertation. A multiscale model of TALL-3D was built by using the coupled ATHLET-ANSYS CFX code system to perform flow simulations, based on fast 3D approaches, and to perform uncertainty and sensitivity analysis of the results. Based on the information about the variability of the input parameters, a series of code simulations were performed for the selected experimental scenario to investigate the uncertainty propagation in the coupled multiscale computational chain, and the simulation results with uncertainty intervals were compared with the experimental data. First, the tolerance limits calculated by the coupled and standalone STH (System Thermal-Hydraulics) codes in the blind phase were used to analyze the physical phenomenon during the transient in terms of mass flow rates and temperature distributions. The improvement in predictive capability compared to the standalone best-estimate simulation, was verified. Then, the lessons learned in the blind phase were summarized to modify the computer model and improve the predictive capability in the open phase. It was found that the best-estimate results provided by the coupled code system, together with the uncertainty intervals (tolerance limits), were capable of covering the experimental data. It is demonstrated that the uncertainty quantification approach can be applied to the coupled code system and is able to provide uncertainty intervals for simulation results, providing more comprehensive information about the real system. Sensitivity analysis allows to determine the influences of input variables on the output/response, which can be used for the dimension reduction and provide a reference for the experimentalists and manufacturers to reduce the system uncertainty. Finally, surrogate models based on several key output parameters were built using the DACE MATLAB package, and the sensitivity analysis was performed based on the best surrogate model.

2. A coupled 3D neutronics/thermal-hydraulics time-dependent model of a single flow channel for local responses: in the SMDFR core, the nuclear chain reaction

and heat transfer occurred simultaneously, and thus a strong coupling effect of the neutronics and thermal-hydraulics fields was expected and thoroughly investigated taking into account of the input uncertainty, under both steady and transient scenarios. The results indicate that this dual fluid design is capable to be implemented in an advanced nuclear system and has a high perturbation tolerance.

3. A coupled Monte-Carlo and 3D computational fluid dynamics model of the entire core for global responses: the entire core of the SMDFR was modeled by coupled Serpent and COMSOL Multiphysics codes, for high-fidelity simulation of the neutronics and thermal-hydraulic behaviors. A 1D system model was then constructed, taking into account of the drift effect of the delayed neutron precursors. The results obtained from the high-fidelity coupled model were used as reference data, and the errors introduced by the dimensionality reduction of the 1D system model were quantified based on the inverse uncertainty propagation technique. In addition, an uncertainty-based control scheme of this system was established and optimized. The methodology of uncertainty-based optimization of the control system was proposed and carried out for two cases, namely, step load change and linear load change, where the uncertainty was quantified using the tolerance limits technique, and the propagated uncertainty was quantitatively considered during the optimization process. The performance of the control system before and after optimization was compared. It should be noted that the developed methodology of uncertainty-based optimization is suitable not only for the design of control systems of nuclear systems, but also applicable to the design of control systems of other energy systems subject to uncertainties. The obtained results indicate that the designed control system was able to maintain the stability of the SMDFR system and to regulate the power as expected. By applying the methodology of the uncertainty-based optimization, the delivered control system was optimal and proved to have the best performance, demonstrating less overshoot, less oscillation, and less time needed to reach the new steady state.

7.2 Conclusion

First, by comparing the numerical results with experimental data of the TALL-3D facility in the context of the SESAME project, the application of the uncertainty quantification and sensitivity analysis methodology in the coupled code system was validated and a deeper and more comprehensive understanding of the system was achieved by this methodology, such as: the uncertainties of the simulation results with a certain confidence level, the most influential parameters in different scenarios. Subsequently, the validated uncertainty quantification and sensitivity analysis methodology was applied to coupled multiphysics models of a real advanced reactor system operating with a similar fluid, in which implicit and explicit coupling schemes were applied. It was demonstrated that the system behavior and the coupling effect of physical fields can be predicted and captured by the coupled codes. Uncertainties of output variables under steady and transient scenarios were quantified, and key parameters were identified through the quantitative sensitivity analysis. Based on these results, it was verified that the SMDFR system is resistant to perturbations and has the ability of self-stabilization. In addition, the weak points of the system were identified and a reference for further development and improvement was provided. Finally, the inverse uncertainty propagation method was applied to the system modeling in MATLAB to account for the loss of precision due to dimensionality reduction. Based on the provided system model with uncertain parameters, a control system was designed and optimized using the uncertainty-based optimization method.

In summary, the application of uncertainty quantification and sensitivity analysis methodology in multiphysics and multiscale coupling approaches is capable of providing more reliable and more informative results compared to the conventional standalone best-estimate approaches. As a result, this work makes a significant contribution to the development, validation and application of various coupling schemes and uncertainty quantification and sensitivity analysis approaches in the modeling of complex nuclear facilities, and provides a reference for further design and safety analysis activities.

7.3   Outlook

Regarding the coupling methodology, the coupling of COMSOL Multiphysics and ATHLET will be studied for the investigation of the entire SMDFR system: not only the reactor core, but also the other components in the primary and secondary circuits. For the reactor core, the addition of fluid-structure interaction and fuel burnup calculations could be considered for a more comprehensive investigation of the long-term behavior of the system.

Although the thermal-hydraulics related to the macroscopic system is the most important objective, further uncertainty quantification and sensitivity analysis of the local part (e.g. the TS) of the system would be helpful to understand the global system behavior. It should be noted that the uncertainty comes not only from the input parameters, but also from the model parameters. However, since none of the codes used in this dissertation are open source, the manipulation of the parameters in the constitutive model of these codes is not feasible, which means that the model uncertainty quantification should be performed based on open source codes (such as OpenFOAM). Future work could also be dedicated to further improve the surrogate modeling by modifying the regression and correlation functions.

Bibliography

[Car16]   Beatrice Carasi. Compressibility options and buoyancy forces for flow simulations. https://www.comsol.com/blogs/compressibility-options-and-buoyancy-forces-for-flow-simulations, Aug 2016. Accessed: 2020-07-30.

[COM20a]   COMSOL. *CFD Module User's Guide*. COMSOL, 2020.

[COM20b]   COMSOL. *Heat Transfer Module User's Guide*. COMSOL, 2020.

[COM20c]   COMSOL. *Introduction to COMSOL Multiphysics*. COMSOL, 2020.

[CSLM17]   Heejin Cho, Aaron Smith, Rogelio Luck, and Pedro J Mago. Transient Uncertainty Analysis in Solar Thermal System Modeling. *Journal of Uncertainty Analysis and Applications*, 5(1):1, 2017.

[DKRC75]   VN Desyatnik, SF Katyshev, SP Raspopin, and Yu F Chervinskii. Density, surface tension, and viscosity of uranium trichloride-sodium chloride melts. *Soviet Atomic Energy*, 39(1):649–651, 1975.

[ea16a]   G. Lerchl et al. *ATHLET 3.1A User's Manual*. GRS-P-1, Schwertnergasse 1, 50667 Köln, Germany, March 2016.

[ea16b]   H. Austregesilo et al. *ATHLET 3.1A Models and Methods*. GRS-P-1, Schwertnergasse 1, 50667 Köln, Germany, March 2016.

[FSA+15]   Concetta Fazio, VP Sobolev, A Aerts, S Gavrilov, K Lambrinou, P Schuurmans, A Gessi, P Agostini, A Ciampichetti, L Martinelli, et al. Handbook on lead-bismuth eutectic alloy and lead properties, materials compatibility, thermal-hydraulics and technologies-2015 edition. Technical report, Organisation for Economic Co-Operation and Development, 2015.

[Gef17]   Clotaire Geffray. *Uncertainty Propagation Applied to Multi-Scale Thermal-Hydraulics Coupled Codes: A Step Towards Validation*. PhD thesis, Technische Universität München, 2017.

[GIF17]   GIF. 2017 gif annual report. Technical report, OECD Nuclear Energy Agency for the Generation IV International Forum, 2017.

[GJK+15]   Dmitry Grishchenko, Marti Jeltsov, Kaspar Kööp, Aram Karbojian, Walter Villanueva, and Pavel Kudinov. The TALL-3D facility design and commissioning tests for validation of coupled STH and CFD codes. *Nuclear Engineering and Design*, 290:144–153, 2015.

© The Editor(s) (if applicable) and The Author(s), under exclusive license to Springer Fachmedien Wiesbaden GmbH, part of Springer Nature 2023
C. Liu, *Multiscale and Multiphysics Modeling of Nuclear Facilities with Coupled Codes and its Uncertainty Quantification and Sensitivity Analysis*,
https://doi.org/10.1007/978-3-658-43422-9

[GKHL+08]   H Glaeser, B Krzykacz-Hausmann, W Luther, S Schwarz, and T Skorek. Meth-
            odenentwicklung und exemplarische anwendungen zur bestimmung der aus-
            sagesicherheit von rechenprogrammergebnissen. *GRS Rep*, 2008.

[Gla08]   Horst Glaeser. GRS method for uncertainty and sensitivity evaluation of code
            results and applications. *Science and Technology of Nuclear Installations*,
            2008, 2008.

[GPL+20]   Dmitry Grishchenko, A Papukchiev, C Liu, C Geffray, M Polidori, K Kööp,
            Marti Jeltsov, and Pavel Kudinov. Tall-3d open and blind benchmark on natural
            circulation instability. *Nuclear Engineering and Design*, 358:110386, 2020.

[HRW+17]   Armin Huke, Götz Ruprecht, Daniel Weißbach, Konrad Czerski, Stephan Got-
            tlieb, Ahmed Hussein, and Fabian Herrmann. Dual-fluid reactor. In *Molten Salt
            Reactors and Thorium Energy*, pages 619–633. Elsevier, 2017.

[Kee65]   G.R. Keepin. *Physics of Nuclear Kinetics*. Addison-Wesley series in nuclear
            science and engineering. Addison-Wesley Publishing Company, 1965.

[KM09]   Kalimuthu Krishnamoorthy and Thomas Mathew. *Statistical tolerance
            regions: theory, applications, and computation*, volume 744. John Wiley &
            Sons, 2009.

[LLLMJ20]   Chunyu Liu, Xiaodong Li, Run Luo, and Rafael Macian-Juan. Thermal
            hydraulics analysis of the distribution zone in small modular dual fluid reactor.
            Metals, 10(8):1065, 2020.

[LNS02]   Søren Nymand Lophaven, Hans Bruun Nielsen, and Jacob Søndergaard.
            DACE: a Matlab kriging toolbox, volume 2. Citeseer, 2002.

[LPMJ20]   Chunyu Liu, Angel Papukchiev, and Rafael Macián-Juan. Uncertainty and
            sensitivity analysis of coupled multiscale simulations in the context of the
            sesame eu-project. *International Journal of Advanced Nuclear Reactor Design
            and Technology*, 2:117–130, 2020.

[LPV+15]   Jaakko Leppänen, Maria Pusa, Tuomas Viitanen, Ville Valtavirta, and Toni
            Kaltiaisenaho. The serpent monte carlo code: Status, development and appli-
            cations in 2013. *Annals of Nuclear Energy*, 82:142–150, 2015. Joint Inter-
            national Conference on Supercomputing in Nuclear Applications and Monte
            Carlo 2013, SNA + MC 2013. Pluri- and Trans-disciplinarity, Towards New
            Modeling and Numerical Simulation Paradigms.

[MCS+07]   J Mahaffy, B Chung, C Song, F Dubois, E Graffard, F Ducros, M Heitsch,
            M Scheuerer, M Henriksson, E Komen, et al. Best practice guidelines for the
            use of cfd in nuclear reactor safety applications. Technical report, Organisation
            for Economic Co-Operation and Development, 2007.

[Men94]   Florian R Menter. Two-equation eddy-viscosity turbulence models for engi-
            neering applications. *AIAA journal*, 32(8):1598–1605, 1994.

[Mor91]   Max D Morris. Factorial sampling plans for preliminary computational exper-
            iments. *Technometrics*, 33(2):161–174, 1991.

[NMH+97]   Ove Nilsson, Harald Mehling, Rainer Horn, Jochen Fricke, Rainer Hofmann,
            S Mueller, Robert Eckstein, and Dieter Hofmann. Determination of the thermal
            diffusivity and conductivity of monocrystalline silicon carbide (300–2300 K).
            High Temperatures-High Pressures, 29:73, 01 1997.

[Noe67]   Gottfried Emanuel Noether. *Elements of nonparametric statistics*. Wiley, 1967.

[O'H06]   Anthony O'Hagan. Bayesian analysis of computer code outputs: A tutorial. *Reliability Engineering & System Safety*, 91(10–11):1290–1300, 2006.

[Pap16]   Angel Papukchiev. Influence of solid structure and conjugate heat transfer modeling on the liquid temperature distribution inside a cylindrical test section. In *Proc. of ICAPP 2016 conference*, April 2016.

[PGGK16]   A. Papukchiev, C. Geffray, D. Grishchenko, and P. Kudinov. Application and Validation of the Multiscale Code ATHLET-ANSYS CFX for Transient Flows in Next Generation Reactors. In *OECD/NEA CFD4NRS-6 Workshop*, September 2016.

[PGL+19]   Angel Papukchiev, Clotaire Geffray, Chunyu Liu, Dmitry Grishchenko, and Pavel Kudinov. Code validation activities within the European SESAME project and some lessons learned. In *SESAME International Workshop*, March 2019.

[PLW+11]   A. Papukchiev, G. Lerchl, J. Weis, M. Scheuerer, and H. Austregesilo. Development of a coupled 1D-3D thermal-hydraulic code for nuclear power plant simulation and its application to a pressurized thermal shock scenario in PWR. In *NURETH-14*, September 2011.

[PMMM09]   Mohammad Pourgol-Mohamad, Mohammad Modarres, and Ali Mosleh. Integrated methodology for thermal-hydraulic code uncertainty analysis with application. *Nuclear technology*, 165(3):333–359, 2009.

[PMMM10]   Mohammad Pourgol-Mohamad, Ali Mosleh, and Mohammad Modarres. Methodology for the use of experimental data to enhance model output uncertainty assessment in thermal hydraulics codes. *Reliability Engineering & System Safety*, 95(2):77–86, 2010.

[RBSB70]   RC Robertson, RB Briggs, OL Smith, and ES Bettis. Two-fluid molten-salt breeder reactor design study (status as of january 1, 1968). Technical report, Oak Ridge National Lab., 1970.

[SAA+10]   Andrea Saltelli, Paola Annoni, Ivano Azzini, Francesca Campolongo, Marco Ratto, and Stefano Tarantola. Variance based sensitivity analysis of model output. design and estimator for the total sensitivity index. *Computer Physics Communications*, 181(2):259–270, 2010.

[SES]   SESAME. Sesame h2020 project. http://sesame-h2020.eu/.

[SLW+18]   Vikram Singh, Matthew R. Lish, Alexander M. Wheeler, Ondřej Chvála, and Belle R. Upadhyaya. Dynamic modeling and performance analysis of a two-fluid molten-salt breeder reactor system. *Nuclear Technology*, 202(1):15–38, 2018.

[Sob67]   Ilya M. Sobol. Distribution of points in a cube and approximate evaluation of integrals. *U.S.S.R Comput. Maths. Math. Phys. 7*, pages 86–112, 1967.

[Sob01]   Ilya M Sobol. Global sensitivity indices for nonlinear mathematical models and their monte carlo estimates. *Mathematics and computers in simulation*, 55(1–3):271–280, 2001.

[TL74]   Mieczyslaw Taube and J Ligou. Molten plutonium chlorides fast breeder reactor cooled by molten uranium chloride. *Annals of Nuclear Science and Engineering*, 1(4):277–281, 1974.

[Wan17]   Xiang Wang. *Analysis and evaluation of the dual fluid reactor concept*. PhD thesis, Universitätsbibliothek der TU München, 2017.

[Wic18]   Damar Canggih Wicaksono. *Bayesian Uncertainty Quantification of Physical Models in Thermal-Hydraulics System Codes*. PhD thesis, Ecole Polytechnique Fédérale de Lausanne, 2018.

[Wil98]   Christopher KI Williams. Prediction with gaussian processes: From linear regression to linear prediction and beyond. In *Learning in graphical models*, pages 599–621. Springer, 1998.

[WK14]   Xu Wu and Tomasz Kozlowski. Uncertainty quantification for coupled Monte Carlo and thermal-hydraulics codes. In *Transactions of American Nuclear Society*, 2014.

[WLMJ18]   X. Wang, C. Liu, and R. Macián-Juan. Dynamics and stability analysis of dft using u-pu and tru fuel salts. In *Proceedings of the 2018 International Congress on Advances in Nuclear Power Plants, ICAPP 2018*, 2018.

[WLMJ19]   Xiang Wang, Chunyu Liu, and Rafael Macian-Juan. Preliminary hydraulic analysis of the distribution zone in the dual fluid reactor concept. *Progress in Nuclear Energy*, 110:364–373, 2019.

[WMJ17a]   Xiang Wang and Rafael Macian-Juan. Comparative study of basic reactor physics of the dfr concept using u-pu and tru fuel salts. In *International Conference on Nuclear Engineering*, volume 57830, page V005T05A001. American Society of Mechanical Engineers, 2017.

[WMJ17b]   Xiang Wang and Rafael Macian-Juan. Comparative study of thermal-hydraulic behavior of the dfr using u-pu and tru salt fuels. In *International Conference on Nuclear Engineering*, volume 57847, page V006T08A004. American Society of Mechanical Engineers, 2017.

[WMJ18]   Xiang Wang and Rafael Macian-Juan. Steady-state reactor physics of the dual fluid reactor concept. *International Journal of Energy Research*, 42(14):4313–4334, 2018.

[WMJS15]   X Wang, R Macian-Juan, and M Seidl. Preliminary analysis of basic reactor physics of the dual fluid reactor-15270. In *ICAPP 2015 Proceedings*. Societe Francaise d'Energie Nucleaire (SFEN), 2015.

Printed in the United States
by Baker & Taylor Publisher Services